The publisher and the University of California Press Foundation gratefully acknowledge the generous support of the Atkinson Family Foundation Imprint in Higher Education.

Championing Science

₁nce

Communicating Your Ideas
to Decision Makers

Roger D. Aines
and Amy L. Aines

UNIVERSITY OF CALIFORNIA PRESS

University of California Press, one of the most distinguished university presses in the United States, enriches
lives around the world by advancing scholarship in the
humanities, social sciences, and natural sciences. Its
activities are supported by the UC Press Foundation and
by philanthropic contributions from individuals and
institutions. For more information, visit www.ucpress.edu.

University of California Press
Oakland, California

Library of Congress Cataloging-in-Publication Data

Names: Aines, Roger D., 1958– author. | Aines, Amy L.,
    1956– author.
Title: Championing science : communicating your ideas to
    decision makers / Roger D. Aines and Amy L. Aines.
Description: Oakland, California : University of
    California Press, [2019] | Includes bibliographical
    references and index. |
Identifiers: LCCN 2018033185 (print) | LCCN 2018038538
    (ebook) | ISBN 9780520970182 (ebook) |
    ISBN 9780520298071 (cloth : alk. paper) |
    ISBN 9780520298095 (pbk. : alk. paper)
Subjects: LCSH: Communication in science. | Business
    communication. | Business presentations. | Persuasion
    (Psychology)
Classification: LCC Q223 (ebook) | LCC Q223 .A35 2019
    (print) | DDC 501/.4—dc23
LC record available at https://lccn.loc.gov/2018033185

Manufactured in the United States of America

27 26 25 24 23 22 21 20 19
10 9 8 7 6 5 4 3 2 1

Roger:
To my mother, Ann Aines, who taught me to love all aspects of science

Amy:
To Kathie Wiener, who left us too soon, and who taught me that words matter

Contents

Illustrations

TABLES

Preface

Scientists are great communicators—with other scientists. We are schooled in the exacting art of talking to our professors and colleagues, people deeply steeped in both the importance and the nuance of our topic. We can talk about the incredible details of modern science in an efficient way, condensing complex arguments into short discourses. But once we go out beyond the academic world to make an impact, scientists from every discipline face a brand new challenge—communicating science to decision makers.

Decisions about which scientific endeavors are advanced and how they are pursued usually get made by people who are not experts in the field. Corporate chief technology officers, elected officials, government program managers, venture capitalists, heads of nongovernmental agencies, and, often, senior management have the power to award funding and support new discoveries. These decision makers are well educated, hardworking, sincere, and extremely busy. Over the course of a day, they may be expected to make important decisions on topics spanning a myriad of unrelated fields. It is incumbent on us as scientists to quickly and effectively make our case. We must learn to talk about our work in succinct and compelling ways that convince the people who are pivotal to our success to take action.

From my work as a scientist and a manager of large-scale projects at a major national laboratory, I am acutely aware that most scientists could use a crash course on how to champion their science. Amy has dedicated her career to teaching corporate executives how to communicate effectively with their employees, investors, customers, and the public. She's helped her share of scientists and technologists create powerful messages and layer technical information in a compelling, understandable, and actionable way. Her motto is "Words Matter." When I met her, I realized that she would be the perfect coauthor (we ultimately married, but that is a different story!).

Throughout this book, I offer the perspective of a scientist. Amy provides the nuts and bolts of how to improve communication. You will quickly note the juxtaposition of her clear and direct style with my more baroque inclinations. The collaboration brings our distinctly different experience and interests to bear as we give stories and examples from our respective careers. We share the belief that it is now more important than ever for scientists to become better communicators.

The Internet has introduced a new challenge for talking about science. In a world where simply saying something online has impact—whether it's true or not—scientists compete not only with other science and other belief systems but also with deliberate falsification. We must convince decision makers that the information we are providing is both important and correct. They must trust us. This gets complicated when decision makers are overwhelmed with information from multiple sources and can't begin to parse what they should believe. The "what you should eat to be healthy" problem in the popular press is a dramatic example. Scientists eager to get their research into the public eye make limited or even flawed studies seem to carry the same weight as multiyear studies with thousands of participants. The public only sees that scientists seem to say different things on different days. And so, rather than try to work out what is right or wrong, people often make decisions based on what fits their worldview. Investing the time to shape that view of the world and build a trusted relationship is central to becoming a champion for your science.

To convince today's decision makers, you have to establish your credibility and make the case for why the problem you are addressing matters *before* you can talk about the opportunity and ask for support. Constructing and presenting your case must be done masterfully and efficiently. During our research for this book, we interviewed people who can help you do this by providing their voice of experience and examples of effective science communication. Relevant excerpts of these interviews are woven throughout the book. The full interviews can be found at www.championingscience.com.

Amy and I wrote this book to teach science students and professionals how to use words and visuals to compel action. We want to arm scientists with proven techniques for communicating more powerfully so they can continue to change the world for the better.

Roger Aines
Livermore, California

Acknowledgments

We are beholden to Steve Bohlen, Jay Davis, Nancy Floyd, Dick Friedman, Julio Friedmann, Rob Socolow, Josh White, Bill Young, and several Washington sources who chose to remain unnamed but agreed to be interviewed. Their real-world experience and advice provides invaluable perspective to our readers and brings this book to life. We also appreciate the input we received from Matt Stepp and Megan Nicholson, formerly at the Center for Clean Energy Innovation, who we met with early on and who provided insight into how science policy is influenced.

Roger could not have cowritten this book without the decades of experience that Lawrence Livermore National Laboratory and the U.S. Department of Energy heaped upon him. He is grateful for the excellent images and visual examples provided by Josh White, Josh Stolaroff, and Congwang Ye and appreciates the many colleagues who encouraged this aspiration, including Tarabay Antoun, Jeff Roberts, and Doug Rotman.

We are especially thankful that our dear friend, the author Ayelet Baron, convinced us that we had valuable insight to share with the world and that we could write a book as the vehicle for that. Our research was enriched by our archeologist son Ethan Aines, who recommended both Thomas Kuhn's *Structure of Scientific Revolutions* and Marcel Mauss's *The Gift* as sources that informed significant portions of the book.

We are grateful for the support from Cindy Akins-Duffin, who, while on assignment in the Office of Science and Technology Policy, arranged a roundtable discussion and a number of interviews with staff and science communicators from several agencies. Their input and perspective reinforced the need to address this topic, and one of them coined a phrase that became a major theme for the book—*self-aware and self-correcting.*

Our editor Lyn Uhl shared our vision and broadened it by encouraging us to consider what students need to be successful in their future careers. She taught us the power of pedagogy, provided well-timed and wise guidance, and solicited input from four reviewers: Scott St. George, Assistant Professor of Geography at the University of Minnesota; Jennifer Wilcox, the James H. Manning Chaired Professor of Chemical Engineering at Worcester Polytechnic Institute; Erika Check Hayden, San Francisco-based journalist and director of the Science Communication Program at the University of California, Santa Cruz; and Jane Dunphy, Senior Lecturer in MIT's Department of Global Studies and Languages.

The early feedback helped immensely, resulting in the addition of sections on information deficit thinking and managing emotions. The second set of reviews challenged us to rethink the subtitle and structure of the book, which now has four parts and two fewer chapters. Lyn also introduced us to Lisa Moore, who provided early editorial guidance on the structure of the book, which led to further worthwhile refinements. Recommendations from UC Press editor Dore Brown and copyeditor Genevieve Thurston brought considerable clarity to the text.

We owe a particular debt of gratitude to Dr. George Chang, a world-renowned surgeon at MD Anderson, and his terrific team, who literally saved Roger's life. We would be remiss not to mention e-Harmony, which brought us the opportunity to collaborate as authors and become life partners.

Finally, we are grateful for the long hours of meetings and conferences in both of our careers that inspired us to think that we had something useful to add to scientist's communication skills.

PART ONE

Fundamental Concepts for Championing Science

1

Becoming a Champion

Robert Millikan had a problem. In the early 1920s, the established titans of California learning, Stanford and the University of California, Berkeley, had thwarted his attempt to put his new university on a solid footing through state funding. Millikan was chairman of the Executive Council of the California Institute of Technology (Caltech), and his plan was to aggressively recruit top-notch scientists to launch Caltech into the first rank of U.S. schools. Experimental physics, Millikan's specialty, was to be the centerpiece of that effort. George Hale, Millikan's collaborator in this plan, was drawing attention with his bold new science using the Mt. Wilson telescope, one of the largest in the world. Hale was sure that he could surpass that one-hundred-inch instrument, which he had made for the Carnegie Institute, by building a record-shattering two-hundred-inch telescope. His article in the *Atlantic Monthly* extolling the scientific discoveries that would be possible with such a behemoth captured the imagination of a public just discovering science—and the post–World War I growth of American education guaranteed that many of the magazine's readers would have the knowledge to follow and discuss such advanced topics.[1]

Millikan was a brilliant scientist. In 1923, he received the Nobel Prize in Physics for measuring the charge on the electron. But perhaps his greatest talent was as an organizer and champion for science. He knew that Hale could change the world with his huge new telescope, and he knew that many of his faculty could match that accomplishment.

Ultimately, three of the original Caltech professors took home the Nobel Prize. As he looked around his home in Pasadena, he knew that the wealth of Southern California, launched by gold, solidified by railroads, and now being multiplied by oil, movies, and airplanes, was clearly enough to support a major university. How could he inspire the investments that this wealth could support?

The answer, as often happens, lay in dinner and discussion. The new faculty of Caltech met regularly for dinner at a traveling table of twelve. The discussions, and responses from friends occasionally invited to fill the table, suggested to Millikan that the way to encourage interest, and ultimately investment, in science was to invite wealthy residents of Pasadena to dinner. Let them sit in on the excitement and share in the anticipation of new problems and approaches to solving them, and in the course of that intellectual thrill they would realize that they held the key—the funds—to make it happen.

With a fortunately timed $500,000 donation of stocks that Caltech sold just before the crash of '29, Millikan was able to build the Athenaeum, advertised as a combined faculty and private dining club for Caltech; the Huntington Library, whose founders built the transcontinental railroad; and the Mt. Wilson Observatory. His regular dinners with the Pasadena elite enabled Millikan to fund his new institution without the help of the State of California, despite the desire of Stanford and Berkeley to keep Southern California firmly out of the academic limelight.

By 1936, the institute had established itself as a center for aeronautical (and soon space) science and engineering. Hale and Millikan were able to build the two-hundred-inch Mt. Palomar telescope—the largest telescope in the world for nearly half a century, and still one of the great scientific instruments—without any government funding by convincing the Rockefeller Foundation to finance the project. The discovery of the positron (Carl Anderson), the elucidation of the nature of the chemical bond (Linus Pauling), and the creation of modern seismology (Charles Richter) were all made possible not because the United States Congress felt that the military or the economy needed some scientific groundwork laid but because Millikan was able to share the excitement and promise of new science with wealthy individuals who were inspired by this real-world Buck Rogers. This champion of science would no doubt be just as adept at convincing a politician or some federal agency if he were alive today. His ability to involve others in the thrill of scientific discovery serves as a consummate example of how scientists can bring our excitement and our breakthroughs to key decision makers.

MISSILES, RADAR, AND NUCLEAR DETERRENCE: THE BEGINNING OF BIG GOVERNMENT SCIENCE

Up until the late 1930s, big science remained the bailiwick of the Robert Millikans of the world. Government made sure that agriculture, mining, and the technologies of the industrial revolution got the educational attention they required for safe and somewhat sustainable resource utilization. The land-grant schools taught us how to keep our soil from blowing away. But the physics breakthroughs of the beginning of the twentieth century were made largely without the support of the world's governments. World War II taught us another lesson: big science led to incredible advances, which ultimately translated into better lives at home—at least once we recovered from the destructive power of advanced weapons.

Suddenly, government was in the science business, led by brilliant minds like Vannevar Bush, who spearheaded U.S. research during World War II and helped create the National Science Foundation with his report *Science, the Endless Frontier,* which was the beginning of a new approach to government science not just in the United States but around the world.[2] The new mantra was "Government support of science equals better lives for all." Without question, that has been true from the eradication of polio to the economic advances of a world where, now, for the first time in history, the majority of the world's people soon will be middle class.[3] (For more on Vannevar Bush, see chapter 16, "High-Impact Examples of Championing for a Cause.")

During the postwar period, the enormous expenditures of the National Institutes of Health, the National Science Foundation, the U.S. Department of Energy, and their equivalents around the world changed the face of science. Science was now a larger fraction of the world's investments. With this came the need to ensure that the torrent of money created things of value, and that led to formal processes for funding—calls for proposals, peer reviews, funding panels, and agency oversight.

The burden of excitement shifted in this world. Now, the scientist was more responsible for making sure that other scientists thought his or her ideas and practices were solid and defensible. The job of being champions for science fell to politicians, cabinet secretaries, and industry leaders. The president himself was called on to champion the U.S. charge into space in 1961, and a massive bureaucracy was created not only to justify and enliven the idea of space science but also to see it executed in an effective, safe, and, when university or industrial research funding was required, fair manner.

So the task of championing science was taken up by leaders who were often just as interested in creating a boon for their district as in making the world a better place. Scientists were trained to be eminently accurate, fair, and honest—to the betterment of the overall scientific enterprise, which needed a huge cadre of reliable workers to advance the great science of the age. The benefits to humanity were clear. From those dependable hordes of scientists came the transistor, giant growth in food production, and continuous increases in longevity worldwide. Scientists happily let others drive the agenda and champion the need for science and knowledge.

THE NEW AGE OF SCIENCE CHAMPIONSHIP

Many scientists continue to be content living in the world of science agency proposals and careers built on hundreds of publications in their field. Others no longer find this path financially or intellectually satisfying. Venture-funded technology, the demands of dealing with difficult diseases, and the challenges of bringing the benefits of developed nations to the entire world while keeping our climate in check mean we need to move beyond the postwar model of science funded only by giant government programs.

Science is, and must be, much more than providing the best response to the latest call for proposals or, for more mature and influential scientists, participating in writing the text for that call. Science once again needs champions who are scientists—champions who will make the pitch for funding new inventions or advancing the fight against hazards that impact humanity. If scientists can explain why science matters with the accuracy, enthusiasm, promise, and capability of people like Millikan, they can compel decision makers to provide support.

Certainly there is self-interest in this activity. Scientists love science. They love to be paid to practice science and love to have the funding to make incredible scientific advances. But just as Robert Millikan did, scientists must elevate their efforts to bring greater benefits to the world. Millikan's passion attracted a faculty who were willing to risk oblivion by leaving established positions on the East Coast, San Francisco, and Europe for the orange groves of Pasadena. He brought his enthusiasm to bear on physics, aeronautics, and the exploration of the universe. Along the way, he built a considerable university, but he never made that his primary goal—it was just one of the things that would help make the science happen. For promoters of science, that is one of the great challenges. *Successful advocacy without excessive self-interest equals championship.*

But responding to calls for proposals that are intended to compare the ideas and capabilities of scientists who are very much alike will never be adequate for this new age. And yet that is what scientists are trained to do in graduate school. This book is intended to help increase any scientist's ability to make the case for excitement and bring decision makers and colleagues along to become a part of doing great things. Our goal is to unleash a little of Robert Millikan's spirit in all of us. We can be better at convincing our sponsors, our management, and our collaborators that there is much, much more to be accomplished. We can all be champions of science.

TALKING TO DECISION MAKERS

Robert Millikan spent more than a decade building a new institution, drawing together a community of like-minded scientists, gaining support from new sources, and creating scientific advances on a scale and level of impact much larger than what he saw in the other universities of his day.

While we hope that this book inspires some of you to take on enormous scientific crusades, you are more likely to pull it from the shelf for the common task of preparing to influence someone who can help you advance science that you believe is important. This book is not about communicating science to the general public. It's about motivating the decision makers and colleagues who can help support, fund, and implement your work. We think you will find our advice useful when talking to a program manager at a science agency, but we see greater value in helping you talk to less technical audiences, like the leaders of your organization, philanthropists, investors, industry partners, politicians, or government planners. The concepts in *Championing Science* will also help you communicate effectively with colleagues in your own field or from different disciplines.

This book is a comprehensive guide to helping students and scientists build vital communication, influence, and emotional intelligence skills. We introduce and build on a five slide approach as a foundation for succinct communication. Through instruction, interviews, and examples, you will learn that inspiring decision makers to act requires extracting the essence, crafting key messages, simplifying visuals, bridging paradigm gaps, and creating compelling narratives. In the last chapter, we reinforce our principles for championing science through the story of Vannevar Bush's transformation of the relationship between government and scientists, the first medical application of accelerator mass spectrometry, and the arranged marriage between Caltech and MIT, which led to the discovery of gravity waves.

We have structured *Championing Science* in four parts:

1. Fundamental Concepts for Championing Science

2. Mechanics of Championing Science

3. Honing Your Communication, Influence, and Emotional Intelligence Skills

4. Applying the Championing Science Skills

Rather than take a sequential walk through the subject, we highlight the most important aspects of championing science first and then add more detail and examples to help you understand and master these principles.

What's at stake if you don't become a better communicator? The scientific contribution you are poised to make could be left to someone else or, worse, never happen at all. Your ideas are always in competition, and not just with other scientific developments but also with all the other things that decision makers could support with their money and influence. We want to help you carry your ideas across the finish line.

ELEVEN TENETS OF CHAMPIONING SCIENCE

What are the basic actions you should take to make your science understandable and compelling? We know Robert Millikan only from historical records, but we think he would agree that there is a basic set of principals—tenets—that are at the core of being an honest and influential champion. Executed well, these eleven actions can help every scientist communicate ideas to change the world. Throughout the book, we develop these concepts in detail, but if you get no further than absorbing this list, you will be on your way to becoming a more effective communicator and science champion.

1. **Be passionate.** Palpable enthusiasm is contagious. It will carry people along for the great ride of science. Sharing what inspires you about your work will help others see its potential.

2. **Build the big picture first.** Resist the temptation to dive into the details. Frame what you say by succinctly explaining what exists today, the future possibilities, and how your work will fill the gap.

3. **Know who's listening.** Think carefully about what your audience knows and their prevailing sentiment. Determine what you want them to think, do, and feel after they hear from you. Find out how they like to receive information and adapt accordingly.

4. **Spend more time on why it matters and less time on how you do it.** Never promote science for the mere sake of science. Always demonstrate the value to people and the planet we inhabit.

5. **Extract the essence.** Formulate your overarching messages and support points. Tell that story. Never dumb it down.

6. **Be understandable.** Use plain, common language. Avoid or translate acronyms. Start from where your audience is, not from where you are. Use iconic references to anchor scientific concepts to everyday, familiar experiences.

7. **Balance precision with impact.** Choose language carefully to be clear and directionally accurate. Long phrases bog down the listener. Think and speak in short sentences. There is no need for hype. Learn to deliver a compelling narrative.

8. **Be human *and* credible.** The integrity of your word must be unquestionable. Verify your facts. Evaluate your sources. Be yourself. Make an emotional connection by showing up as a person first and a scientist second.

9. **Influence patiently.** Convincing decision makers is a process, not a single act of persuasion. Use information as a gift. Engage often to build understanding and show the value of supporting your science. Learn what matters to your audience.

10. **Collaborate thoughtfully.** Advancing your ideas doesn't mean you have to go it alone. Seek out advisors, influencers, and partners who can help carry your science further.

11. **Enable your listeners to act.** Know the purpose of your communication. Make the ask every time. Leverage each conversation and presentation to build support for advancing your work. Remember that you are ultimately building relationships for the long run.

2

Self-Aware and Self-Correcting

The Key to Effective Communication

Great communicators are *self-aware and self-correcting*. We first heard this from a White House staffer describing how she learned to discuss astronomy in the classroom. When she was teaching fifth graders about spectroscopy, she watched their eyes glaze over when she talked about wavelength. She needed to use a term they could relate to: color. In her words, "You learn this by being self-aware and self-correcting, paying attention to how it sounds and how you are getting your message across, and learning from twenty attempts, how to do it right."[1]

Her insight was easy to map onto the experience of other great science communicators. Each one recognizes that, first and foremost, it is their responsibility to make their message understandable and interesting to the listener. The mechanics of presentation and discussion are incredibly important, but by themselves they are not enough to make you a truly effective champion of science. Before we dive into how to present your science, we want to talk about how you view your own science, and yourself.

SELF-AWARENESS

What does it mean to be self-aware? In a communication context, it's the ability to assess whether your listeners are engaged and whether they come away with the information you intended to convey. More broadly, self-awareness is about understanding your self-image and your needs, desires, failings, habits, and everything else that makes you tick. The more you know about yourself, the better you can adapt to suit a situation. The more self-aware you are, the greater your ability to understand your impact on listeners.

As Northwestern University psychology professor Dan McAdams explains, "The stories we tell ourselves about our lives don't just shape our personalities—they are our personalities."[2] How can you gain self-awareness? Reflect on and answer the following questions:

How do you see yourself professionally?

What words would you use to describe yourself?

How accomplished do you feel?

What is your academic or career success story, and how do you live it?

How much ego do you bring to the way you communicate?

Are you humble, or do you act like you know it all?

Are you willing to listen to others' ideas, or do you just push your own?

Do you really listen and seek to understand?

How do people react to the way you share information?

Contemplating these questions will help you begin to see yourself more clearly and understand how others may see you and your actions.

Psychologists talk about the notion of *trial identification.* Essentially, this entails considering what you are about to say, anticipating how it is likely to be interpreted by the listener based on what you know about them, and then choosing your words accordingly. Patrick Casement, a retired analytic psychotherapist and fellow of the British Psychoanalytic Society who has written extensively on communicating effectively in such award-winning books as *Learning from Our Mistakes* and his classic *Learning from the Patient,* advises that being self-aware is more than just a *think before you speak* habit; rather, it is a strategic communications habit.[3] It also sets up the opportunity for structured self-reflection. By anticipating a certain response, you have a frame of reference for considering why you got that response or why you didn't. You will need to pay close attention to both what your listener says and does. For example, be aware of body language. Does the listener break eye contact, push back in their chair, or cross their arms? These nonverbal communication cues are usually strong indicators that your listener is not favorably inclined toward what you have said.

Self-Awareness: Watching Nonverbal Cues

You become a better communicator by being *self-aware and self-correcting.* That starts with paying close attention to and reflecting on how you come across to others. Watch the faces of your listeners. Their eyes and level of engagement with you show whether they are tracking what you are saying. People often nod in agreement when they understand. They make eye contact instead of looking away. Sometimes they will smile when they see the value of the information you are sharing. When you see someone narrow their eyes or appear to squint, it is often a sign of confusion or disagreement. If they suddenly break eye contact after having appeared engaged, it might indicate that something you just said didn't land the way you intended. The questions they ask will also help you discern whether your message was received and understood as you anticipated. Being self-correcting entails learning—from multiple attempts—how to phrase your ideas in the best way. This practice challenges your ego and breeds humility while training you to interpret nonverbal cues so you can adjust your message for maximum impact.

HOW TO BECOME MORE SELF-AWARE

The first step toward developing greater self-awareness is to recognize when true communication—speaking, listening, and understanding—has *not* occurred. Think about situations that did not go as you anticipated. Perhaps you gave a talk where no one asked a question. Or you pitched a project that was flatly rejected. Maybe you presented an approach and a colleague didn't agree with your position. Or you asked your boss to give you time and budget to further your research and got neither. In each case, ask yourself, "Do I know what went wrong?" If you have no idea, it may be because you were so wrapped up in conveying your message that you were not paying attention to body language and other reaction cues or because you had gauged your audience so badly that there was simply no communication. Often, busily formulating what to say next and not really listening causes failure.

When it comes to developing self-awareness, you don't have to go it alone. Ask for honest feedback from someone you trust who observed the situation. Or talk to a close colleague about what transpired and solicit his or her opinion. Receiving feedback is hard, so do your best to show you are truly open to it. Explain why you care and express your desire to improve. Ask for specific examples to help you see yourself the way others do. The more open you are to getting this kind of input, the more rapidly you can improve your self-awareness. Armed with actionable feedback, you can experiment with new approaches and see how they work. This is what it takes to be self-correcting.

Mustering the courage to politely ask the decision maker can also be a very effective window into how to be self-correcting, especially if you are new to the work world and the decision maker is your boss. You show you care and want to learn when you respectfully ask for help understanding what you would have needed to do to get a *yes* instead of a *no*. Ask your boss these questions: What kind of research would you like to see me undertake? Was there anything about my proposed topic that I could refine so that you would be willing to support it? Can you help me understand what *would* be of greater benefit to the organization? Was my explanation clear or do you have questions I can answer for you? These kinds of queries start an important conversation that can help you turn an unsuccessful interaction into a successful one.

If you reflect on a failed attempt and conclude that the audience was to blame, think again. Communication is a two-way street. "I did the best job I could under the circumstances, but they just didn't want to understand," is a poor excuse. Scientists who take comfort in that explanation are not likely to be reading this book, since they aren't taking personal responsibility for their communication skills. They may even think they are great communicators already. But all of us occasionally feel like the audience is at fault—when it is more likely that we misjudged the audience's interest, existing attitudes, or background on the topic.

Every time you address a group, there is a built-in opportunity to gauge your degree of self-awareness and identify ways to be self-correcting in the future. Did you take time to ask questions of your listeners to determine whether they understood you? If you were in broadcast mode for the entire interaction, chances are you were not fully aware of your audience's reaction to you. All of us struggle to place ourselves in our listener's shoes. As you read this book, keep a (virtual!) mirror handy to reflect on your own approaches and biases.

You can become self-correcting by systematically applying the principles we cover. It is a process of experimentation, reflection, seeking and incorporating feedback, and repeating these steps to continue to refine your approach.

3

Extracting the Essence

When we try to help others think the way we do about our science, we need to be clear. Science is complicated—our descriptions of it don't need to be. But since we are scientists, not salesmen, we want people to believe what we say *because they understand*. This is the challenge for a scientist talking to a decision maker. How can you describe your science so that someone without detailed training can both understand and act on it?

Extracting the essence—learning how to find your most important material without ever dumbing it down—is a major theme of this book. Scientists are rarely taught to do this in school, where providing more and more detail is usually the path to success. Outside the classroom, you all too often get only a few minutes to introduce your ideas to a decision maker.

To extract the essence, ask yourself: What background information does my listener need to understand? Is there a way to simplify my explanation? How can I paint a high-level picture of my work that showcases its merits and helps my listener draw the conclusion I want them to after they hear from me?

An important part of this process is identifying your key messages—the main and most important ideas that you want to convey and have people remember. Distill your key messages into well-constructed, carefully chosen phrases that extract the essence of your arguments and bring your science to life with clear and compelling language. If you don't develop your key messages, you are likely to shower people with detail and information that doesn't have a clearly stated, "So what, why should I care?" orientation.

Garr Reynolds is a well-respected author and advisor on how to construct impactful presentations. Garr counsels us to start preparing our content with a blank whiteboard, or a blank sheet of paper—never by sorting through existing slides. He advises us to take the time to outline the major messages we want to get across first, before compiling or creating slides:

> One reason many presentations are so ineffective is that people today just do not take—or do not have—enough time to step back and really assess what is important and what is not. They often fail to bring anything unique, creative, or new to the presentation. This is not because they are not smart or creative beings, but because they did not have the time alone to slow down and contemplate the problem.[1]

Nowhere is the concept of extracting the essence—and focusing on your key messages—more important than in the first few minutes of a discussion or presentation. Short conversations with your colleagues, friends, bosses, program leaders, or sponsors are essential to advancing your prospect of making an impact or getting a new idea adopted. A five-minute discussion, the first page of a proposal or brief slide presentation should generate interest in spending another half an hour on the details.

The brief slide presentation is perhaps the most difficult to create because we are naturally inclined to include more details of our analysis than time or audience attention will permit. It is a daunting prospect to condense the thirty or forty slides you have already prepared. But editing your material to extract the essence is the foundation of championing science.

We call our recommended structure for these short discussions the five slide approach. It can stand alone or function as the setup to a longer conversation. It is a succinct framework for organizing content, and we return to it throughout the book. Whether it is five slides or five sentences, these are the elements that set the context for, and incorporate, your key messages.

THE FIVE SLIDE APPROACH

The five slide concept originated with Jay Davis, the first director of the Department of Defense's Defense Threat Reduction Agency (DTRA).[2] Jay is a gregarious Texan transplanted first to California and then Washington, DC, with his love of homespun wisdom and simple solutions intact. A physically imposing man who looks like he would be

Extracting the Essence Is Not Dumbing It Down

A common phrase heard around the scientists' water cooler is, "I have to dumb it down for this audience." That speaker, and every other speaker who takes this attitude, *fails as a champion* and will almost certainly fail with that audience, no matter what the goal.

Why? This oft-heard reference to dumbing it down implies that your listener isn't smart enough to understand. It suggests a gross oversimplification of concepts that leaves out details and nuances. The act of dumbing it down, therefore, comes with an air of intellectual superiority and disrespect—two mindsets that won't win favor with decision makers.

When decision makers are not deeply steeped in your branch of science, you need to focus on *extracting the essence*, not dumbing it down. The onus is on you to make the information understandable. Effective persuasion requires that your key messages be tailored to your audience. Focus on the most important ideas that you want to convey and have people remember.

much more comfortable tossing telephone poles while wearing a Scottish kilt than fine-tuning an accelerator, Jay is always thinking about how to make friends and get along with people. In 1991, his experience with custom accelerator design and the physics of nuclear weapons landed him a spot on the inspection teams that were tasked with looking for the Iraqi nuclear material production facilities after the First Iraq War.

After months of futile inspections stymied by deliberate misdirection by the Iraqi military and weapons scientists, Jay's team finally spotted an Iraqi convoy of one hundred trucks transporting the end plates of a large electromagnetic separator across the desert. They gave chase, and despite being fired on in the course of photographing the convoy, they were ultimately able to unveil the full scope of Iraq's attempt to build nuclear weapons using an electromagnetic separation technique that had been abandoned by the West decades before. The Iraqis didn't need an elegant and highly efficient nuclear industry like ones the superpowers had to create thousands of weapons—they just wanted a few, and they didn't have to be anything but deadly, stripped to the essence of mass destruction. Jay's ability to see the heart of the matter made it easy for him to identify the significance of the end plates. He was not looking for the complicated centrifuges of U.S. weapon factories but rather for simple systems that would provide enough distinction among uranium isotopes to make a uranium fission weapon. Jay and fellow inspector David Kay's account of this story in *Physics Today* is an outstanding example of how to tell a riveting and politically appropriate science-based story.[3]

Jay has a passion for particle accelerators, those engines of physics that use electromagnetics to make ions go irrationally fast. He is particularly fond of using them to perform mass spectrometry, a method of measuring the exact mass and number of atoms in a sample at ultralow levels, which he does with his beloved huge, ten-million-volt Tandem Van de Graaff accelerator[4] (Remember the one in the science museum that made your hair stand on end? Think bigger, as in the size of a Midwestern corn silo lying on its side). This type of accelerator now makes it easy for archeologists to measure the age of ancient bones and for doctors to estimate the effects of specific drug interactions in the body by measuring as little as a few thousand atoms at a time.

Early in his career, Roger worked for Jay. Jay was asked to create a new organization at Lawrence Livermore National Laboratory that would combine all the environmental and energy scientists into one large team. This would be the first real experience of science in teams bigger than university research groups for many of these scientists. With almost two hundred scientists on the team, it was important to communicate well. Carrying out the mission to support national needs in the fields of environment and energy required Roger and his colleagues to interact with a diverse set of interested parties and sponsors ranging from environmental activists to former Navy fighter pilots now commanding bases that required dramatic groundwater cleanup.

Jay was adamant that the team be able to prepare five slides at the drop of a hat. The goal was to condense their entire argument, including the ask, into slides that could stand alone if needed. Start with those five. That way, even if you don't get through the entire presentation, you will have covered the important material. And as your listeners become involved in the topic, they will help direct any more detailed discussion in ways that are most useful to them. This is much better than waiting forty-five minutes to get to the key point of your presentation and leaving the audience with no time to evaluate it. This avoids perhaps the most grievous outcome of all: a willing sponsor or an influential policymaker being cut off from helping you by a lack of time. Not on Jay Davis's watch.

Jay emphasized that these are the five things you need to cover:

1. The problem
2. The technical gap

3. How to fill the gap

4. Why you or your team are right for the job

5. The ask

You need the same five elements for a short conversation, the beginning of a presentation, or the first few paragraphs of a white paper. In fact, it is a great idea to discipline yourself so that the first two paragraphs of a white paper match both your five-minute pitch and your first five slides in tone, content, and message (see chapter 13, "Translations, Templates, and White Papers"). This way, you have to create your narrative only once, and you are doing it in the best possible way.

Let's look at what the five slides might contain for a research proposal. This information is as essential when you are trying to attract an outside sponsor as it is when you are asking your boss for internal budget approval.

The Problem

What is missing from the world that both you and your listeners care deeply about? It's good to think about this in two stages: first, the overall context, and second, the gap that keeps us from getting from where we are to where we want to be. That gap is something you will be able to fill with a well-defined research project.

Context: "Climate change is well known to be an issue."

Focused problem: "Predicting the effect of climate change on Navy force utilization in the Pacific has recently been highlighted by PACCOM (Pacific Command)."

This use of both large-scale (context) and project-scale (problem) thinking tells the decision maker that you are aware of the big picture but have the self-discipline to define a project that is clearly in their interest area, with a budget that the funder can supply.

When using slides, start with this, *not* with your title slide or a list of collaborators. For the first twenty seconds, you have the listener's complete attention. Put it to the best possible use. Put the problem in front of them. But don't perseverate on this topic; the next slide is the most important.

The Technical Gap

Now make the connection between the problem and the area of science or technology that could be improved to jump forward to a better position regarding the problem. The *gap* keeps the sponsor from moving to where they want to be.

> Name the specific thing we can't do today because of a gap in knowledge or ability: "We can't predict the number or intensity of Pacific typhoons."

> Link it back to the broad problem statement, but focus on the technical need.

The simple phrase "This gap keeps you from achieving your goal" will get any sponsor's attention. This is the slide where you can talk about the conventional way of thinking about the problem and why the gap exists.

This is also the optimal place to make another statement that will always capture a listener's attention: "What if we could?" For instance, "What if we could skip this hard step?" or, "What if we could apply knowledge from another discipline to solve this tough problem?" An audience, particularly a science and technology audience, will always honor this challenge with another minute of rapt attention. This statement is at the heart of what makes us love science.

Now that you have focused your listeners' attention on a key problem and a potential opening to pursue, you can make the first mention of your contribution.

How to Fill the Gap

What sort of science is needed?

> Imagine the future: "If we could reliably do X, we could fill the technical gap and address the large problem in a new way."

> How will this improve the general state of science and technology? (Even a highly focused sponsor wants to know that their efforts fit into the general progress of knowledge.)

Obviously, in a detailed slide deck this is more than a one-slide topic, but in the five slide version, it really is *only one slide*. Be concise. Trust that the viewers do not need a detailed description of your previous accomplishments and your experimental plan—just enough highlights to make them confident and to let them ask you questions, which they always love to do.

Let your listeners absorb the material slowly; make them want to get the details. Give them the opportunity to love your idea in the time they have available.

Why You or Your Team Are Right for the Job

Among the topics you can choose from to answer this question are:

Why you or your team are poised to be efficient and effective.

Why this problem is well suited to your skills, reputation, and facilities.

What you or your collaborators have done previously in this area.

Emphasizing your collaborators and your past history with the problem is important. Your track record counts. The sponsor wants to know that their money will not be risked on an inexperienced team. If the topic is new to you, talk about your experience in a related field.

The Ask

Now we come to the critical part. You have identified a need, and you have an audience that is connected and interested in solving that need. What is the ultimate goal of your work? Your vision of success sets up the ask. Help your audience understand what you are working to achieve and how you plan to go about doing it. This part of your communication is the ideal opportunity to call on your listeners to become collaborators, funders, or other resources to help you advance your science. Help your audience draw the conclusions you want them to and understand why it matters. Be explicit. What is it you want people to do as a result of listening to you? Close with a concrete call to action.

Jay hammered into his teams that the worst possible thing you can do is waste a decision maker's time by not making an ask. How can they participate? How can they advance their agenda by following your lead? How can they convene other resources also interested in the problem? Even if the ask is nothing more than, "Please let me know what you think of this idea," *make the ask*. Every time.

For research proposals, the ask is usually for money, but you should keep this principle in mind for interactions with any decision maker. Be specific so that they are able to invest in you. Give them the information they need to make a case to their oversight committee or slot you into a research theme in their current portfolio. Give them the ability to act.

> How much are you asking for? Put the amount at the top of the slide to provide context for the scope of your work. Under the title is good. E.g., "$350K/year for three years."
>
> What general types of work will be done (experiments, models, etc.)?
>
> What will you provide that actually fills the gap you defined in slides one and two?

Don't be vague in your verbal description of what you will provide. "A report" or "a model" is inadequate. Instead, call it, "An analysis that compares the results of this method to previous techniques and evaluates the time and resource requirements for implementing this approach," or, "A working prototype computer model that embodies the physics described here."

Be sure to give clear examples of what you will do. Asking for money to evaluate what needs to be done will rarely be well received. However, a midstream decision among two or three options to determine the final path is a great intermediate milestone.

Despite Jay's wise advice, presentations of this sort are often longer than five slides. In chapter 13, we discuss formulaic or templated presentations and how to avoid boring your audience to death with the exact slides they requested. But the five slide approach is always a vital discipline for *starting* a discussion, with our without visuals. Before you go into detail about the technical work or members of your team, give the full five slide context. To move on, you can then say, "Let me talk more about the technical details." With a firm grip on the importance, methods, and overall structure of what you are proposing, the listener will have sufficient context to absorb the particulars of how you will succeed.

Getting the five slide approach right takes practice. We recommend that you pilot your pitch with anyone willing to listen. Make sure it is coming across the way you want, and pay close attention to questions asked that indicate where your listeners were lacking key information. Translate feedback into improvements to your language, flow, and overall content. The five slide presentation is the ultimate place to be self-aware and self-correcting.

Is the right number always five? Of course not. In many instances, these five topics will be the ones you want to focus on, but don't be a slave to the number. The idea is to outline your key messages and your request in a minimalist format that encourages the decision maker to ask for the details. If you force yourself to think, "Five is my target," and you find you need six or seven, you will not go wrong.

EDITING TO EXTRACT THE ESSENCE

Editing your thoughts, your writing, or your slides is absolutely necessary to achieve impact with any audience. The five slide constraint will quickly teach you this, although it may be very frustrating at first. If you want your extraordinary science to be recognized as such, it has to be carefully described. If you are writing for *Science* magazine, you can use the full power of the complex language and principles of your field in your elegant, extended descriptions. However, when you are championing science to a nonspecialist decision maker, keep your description both accurate and simple, or the listener might miss your point.

John Maeda is the president of the Rhode Island School of Design and a computer scientist. He describes the path to simplicity simply: "Simplicity is about subtracting the obvious, and adding the meaningful."[5] This is a remarkably hard distinction for scientists to make because most of their science is obvious (to them) and *all* of it is meaningful. In chapter 5, we take a deep dive into why scientists misread their audiences, but for now we can give this interpretation of Maeda's advice: Make sure you put yourself in the audience's shoes. What do they already know, and what will they consider meaningful?

In a similar vein, Albert Einstein is widely reported to have said, "Everything should be made as simple as possible, but no simpler." This quote was attributed to Einstein by the famous composer Roger Sessions and cited in the *New York Times*.[6] Since it purportedly came from the man who simplified physics to general relativity, it was quickly endorsed as a statement of how to conduct understandable science. And given its popular cachet (it has been cited over one million times on the web, according to Google), it is a statement that many people believe holds true.

Even more interesting is that Einstein didn't actually say it.

Apparently, Sessions applied John Maeda's style of logic to something that he heard Einstein say. The quote that we are *certain* Einstein said, as documented by Alice Calaprice in *The Ultimate Quotable Einstein*,[7] was much more scientific: "It can scarcely be denied that the supreme goal of all theory is to make the irreducible basic elements as simple and as few as possible without having to surrender the adequate representation of a single datum of experience."[8]

That sounds like something scientists would say to their colleagues, but we are sure it would be lost on almost everyone else. And yet we can see the critical elements of the simplified version of the statement clearly exposed in the original. What a marvelous example of the conundrum we face when championing science. Einstein's original, highly accurate, and carefully framed statement fails the impact test. Sessions, however, was able to extract the essence.[9] It was Sessions's summary that caught the attention of the world, not the scientist's detailed original.

Certainly, simplicity in expression makes a message more accessible. But what interests us is why it took a composer to produce one of the most famous quotes about science. Do scientists have to insist on all the accuracy and detail, obscuring the clarity of their message?

CAPTURING THE ESSENCE VISUALLY: CLIMATE WEDGE EXAMPLE

Simplicity is vital when the topic is the enormity of what it will take to address climate change challenges. A case in point is managing the emissions of carbon dioxide. Julio Friedmann is one of the world's leading thinkers on this topic. He helped originate the concept of carbon capture and storage, where carbon dioxide from industry and power plants is captured from the smokestack and then stored underground as a fluid, much like oil. Julio got his bachelor's degree in music and a master's degree in geology from MIT and his PhD in sedimentology, the study of how rocks like sandstone form, from the University of Southern California. He has worked at Exxon, the University of Maryland, and Lawrence Livermore Lab. He also served during the Obama Administration as the Deputy Assistant Secretary of Energy for Clean Coal, and the Principal Deputy Secretary for Fossil Energy.

Julio thinks that podiums are holy places because they give speakers an opportunity to change people's minds and affect the way they think and act in the world. When we asked whom he regarded as a true champion of science, he sang the praises of Rob Socolow, Professor Emeritus of Mechanical and Aerospace Engineering at Princeton. Julio reflected on Rob's strong communication skills:

I think Rob spends every waking moment thinking, "How can I educate, how can I communicate complex ideas to people?" Rob has briefed many, many people on Capitol Hill. He has briefed captains of industry and he has spent his share of time with the people who have access to leaders of countries. He ultimately does not view his job as one of influence. He believes it is most important for him to clarify and elucidate. He wants these busy decision makers to understand the issues. He would say, "I don't run a country, I don't run a Fortune 500 company, I don't know all the other things around all the complexities of these guys' jobs, but if I can communicate my point very clearly and they believe me, they will weave that information into whatever else they do." And in a room full of experts who are busy trying to convey as much detail as possible, Rob will do the opposite. He will elicit as much information as he can from the decision makers in the room, so that he can repackage the details in a form coherent to them—and only then re-communicate it back to the policymakers.[10]

Julio saw Rob give a talk at the World Bank about greenhouse gas emissions using his famous wedge diagram. The wedge diagram is Rob's way of communicating the complexities of climate science in the easiest, simplest way to the widest set of people. He was championing a new way of thinking about the problem. To do so, Rob used several approaches that we recommend. He extracted the essence, really understood his audience, and used the wedge, a familiar shape from geometry, to build a bridge for understanding his recommendation. Julio explained the challenge Rob was facing and how he came up with using the wedge to address it:

> Rob was deliberately trying to push back on the idea that then Secretary of Energy Spencer Abraham put forward, which was that we needed a transformational equivalent to the discovery of electricity to solve climate change. Rob's response to that was, "Poppycock, we just need to spend some money and apply the technology we have." As he tried to package that idea, he realized there was always this bucket of stuff, this bag of rocks of technology and a whole bunch of people trying to champion their respective technology. What was lost in all of this was what's useful and what's materially important. And so Rob, with his colleague Steve Pacala, came up with the wedges.
>
> A climate abatement wedge is just a unit that is 50 years long and represents 25 gigatons of carbon (or about 100 gigatons of carbon dioxide) depicted in the shape of a triangle. Rob said, over 50 years, if you want an emissions trajectory that is flat (and we are currently on a trajectory that is up) then everything between up and flat is some amount of emission that you can break up into bite-size wedges and assign one technology to one wedge. Maybe that's a whole bunch of wind power, or maybe that's a whole bunch of

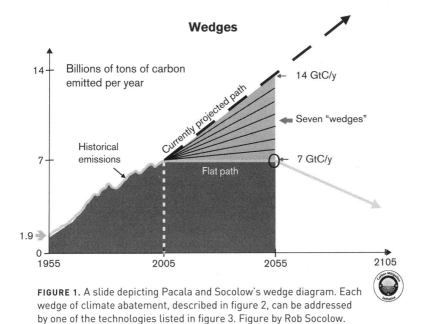

FIGURE 1. A slide depicting Pacala and Socolow's wedge diagram. Each wedge of climate abatement, described in figure 2, can be addressed by one of the technologies listed in figure 3. Figure by Rob Socolow.

energy efficiency for cars, or maybe that's planting a bunch of trees. Whatever it is, it has to be big enough to count. That means, over 50 years, it has to account for that much emissions.

By breaking it down in that very simple way, Rob and Steve were very quickly able to dispense with technologies that were marginal and would never matter. Rob and Steve were also able to say, "We don't care which of these technologies you use as long as you have enough wedges to get the job done." And that allowed people to push forward their own ideas credibly. It also meant that you could very quickly see that you couldn't solve the problem with just one approach, like energy efficiency gains, because you needed seven wedges of that and it just wasn't possible. In a very simple heuristic, the wedges communicated a bunch of very complicated ideas.

Figures 1 and 2 show some of the visuals Rob and Steve created to tell the wedge story. As you will see after you read chapter 9, "Designing Effective Visuals," they are masterful in their ability to convey a singular message by following the best design principles. They adeptly introduce the wedge concept, explain it, and show the elements that could, in combination, be considered technology solutions.

What Is a "Wedge"?

A "wedge" is a strategy to reduce carbon emissions that grows in 50 years from zero to 1.0 GtC/yr. The strategy has already been commercialized at scale somewhere.

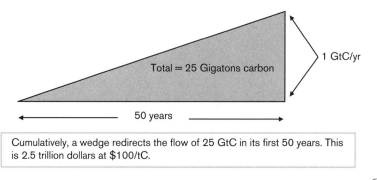

Cumulatively, a wedge redirects the flow of 25 GtC in its first 50 years. This is 2.5 trillion dollars at $100/tC.

A "solution" to the CO_2 problem should provide at least one wedge.

FIGURE 2. Each wedge in figure 1 is a simplified technology implementation pathway that would ultimately reduce emissions by a total of twenty-five billion tons of carbon, or about one hundred billion tons of carbon dioxide. This is another simplification of Rob's—most people think about carbon dioxide, not carbon, so Rob wanted to use different units for different audiences. And rather than multiply the mass by 3.66, the technically accurate number, he multiplied it by 4. Figure by Rob Socolow.

Wedge Strategies Currently Available

The following pages contain descriptions of 15 strategies already available that could be scaled up over the next 50 years to reduce global carbon emissions by 1 billion tons per year, or **one wedge.** They are grouped into four major categories:

Efficiency & Conservation
- Increased transport efficiency
- Reducing miles traveled
- Increased building efficiency
- Increased efficiency of electricity production

Fossil-Fuel Based Strategies
- Fuel switching (coal to gas)
- Fossil-based electricity with carbon capture & storage (CCS)
- Coal synfuels with CCS
- Fossil-based hydrogen fuel with CCS

Nuclear Energy
- Nuclear electricity

Renewables and Biostorage
- Wind-generated electricity
- Solar electricity
- Wind-generated hydogen fuel
- Biofuels
- Forest storage
- Soil storage

FIGURE 3. Possible technologies to make up each of the required seven wedges to reduce carbon emissions to the point that they are no longer growing over time. Figure by Rob Socolow.

Figure 3, from a teacher's guide for a game that uses the wedge concept to teach college students options for climate management, lists a variety of simple options for wedges that extract the essence of the climate technology.[11]

4

Who's Listening?

Know Your Audience

The art of communicating comes down to one essential principle: making sure your message is heard. And that depends on who is listening.

Successful champions invest time in learning as much as they can about the people they are meeting with or addressing. When you conduct a comprehensive audience analysis and understand demographics, expertise, and biases, you can package information to make your conversation more persuasive and effective. Words really matter, especially when the topic is rooted in science.

RESEARCHING YOUR AUDIENCE

Start with the basics:

Who will be in the room?

What are their roles and responsibilities?

What organizations do they represent?

How much are they likely to know about your topic?

Have they already formed an opinion?

Are they likely to have an emotionally based bias?

Do you think they support your premise, or are you about to stand up in front of people who might be naysayers?

Never underestimate the consequences of failing to do sufficient audience research. Roger vividly remembers being "ambushed" at a talk he gave to members of the Bay Area chapter of the American Nuclear Society at a dinner meeting in San Francisco in 2010. It turned out that the chairman of the organization was a climate denier. Unfortunately, Roger had no idea. If he had taken the time to do some homework or asked his audience some questions before the talk, he would have known. Midway through Roger's talk, the chairman could no longer contain himself. He exclaimed, "That's totally ridiculous!" and then proceeded to make his case for why this "climate thing" wasn't real.

You might be thinking that this outcome was inevitable. It wasn't. Yes, the chairman was a naysayer, but if Roger had been aware of that he could have spent some time providing evidence up front to make the case for why climate change was real. And he could have diffused the situation by respectfully acknowledging the chairman's position before informing the group that he was there to provide an update and shed light on the issues. That statement would have prevented what turned out to be a pretty uncomfortable situation for Roger. Lesson learned. Now you know how to avoid making a similar mistake.

The Power of the Internet

If you are meeting someone you know very little about, tap the power of social media, the Internet, and your professional networks to see what additional information you can glean. Following someone on Twitter or LinkedIn can often give you valuable insight into how they think and what's important to them. Following people who work in your area can also help you build relationships that may become important to you in the future. To get background on other scientists, you can search for their publications or see what a Google search turns up. Learn what work has already been done in your area and be sure to give credit where credit is due. The worthy recipient might be sitting in front of you.

Common Connections

When you are addressing a small group of important players or an individual decision maker who can help advance your cause, it's vital to find out as much as you can about them. Yes, this is time consuming, but the rewards can be enormous. You will be surprised how often there is a common link, like school background, that you can use as an icebreaker. It can speed up the time it takes to build rapport and help you focus your discussion on topics of interest while steering clear of points of disagreement. And mentioning something about a person's background is always well received. They appreciate that you took the time to find out more about them.

Giving Credit

At times, you may be speaking to a group of people that includes team members who have contributed to the findings you plan to discuss. Be sure to indicate this clearly when you address your audience. Talk in terms of the work that you and your team have been doing. Don't overlook how much you can spoil relationships and create resistance to getting support for your work by simply failing to share credit with your cocreators. Here's your chance to be humble, likable, and inclusive. Mentioning a collaborator during a talk, as opposed to just listing their name on the title slide, is a meaningful way to show that you value their work.

Audience Phenotypes

What if you are speaking at a conference or work event and have no way of knowing who might show up? You can make some safe assumptions. If you're lucky, you won't encounter too many naysayers. Most likely, you'll be speaking to a mix of these six listener phenotypes:

1. Scientists who are working in a similar field or have an interest in your topic
2. Representatives from companies that see the potential for your science to solve a problem they face or support a new or existing product
3. Industry regulators or government personnel
4. Sponsors, venture capitalists, or investors who could fund your science
5. Peers from your organization
6. Team members working with you

You will find some variation of these phenotypes everywhere you go, even in a poster session. Fortunately, during a small-group discussion you are often able to have casual conversations to gauge your audience. Start by asking about their background and interest in your topic. That will help you figure out the right level of detail and how best to approach the conversation.

Audience Demographics

Asking questions to assess the audience demographics is another technique to help you tailor your presentation to the people who are listening. Questions like, "How many of you are familiar with X?" or, "How many of you attended the earlier session where my colleague spoke?" work well. This technique of engaging your audience helps you be a more effective presenter and helps your listeners connect better to your content. But be careful about doing this in small groups where these questions could make some listeners appear uneducated.

CONNECT WITH YOUR AUDIENCE BY WARMING UP THE ROOM

Once you have successfully scoped out your audience ahead of time, your next task is to connect and set the stage for engaging your listeners. Like a runner stretching before the first lap, you have an invaluable opportunity to warm up the room by chatting casually with a few people before your presentation. If you make a habit of following this advice, you will more quickly engage your listeners when you speak and reap the benefits of establishing early rapport. This helps ease speaker jitters and creates a sense that you are having a conversation with people you know. The more you personalize your opening comments and connect them to the people in the room, the more quickly you'll capture everyone's attention. This approach works just as well when you are in a small group meeting as it does when you are addressing a ballroom full of conference attendees.

Arrive Early

What does it take to warm up the room before you speak at a meeting? Arrive twenty to thirty minutes early. Find your host and make sure he or she has your bio so you can be introduced easily. Having an extra copy with you is always a good idea. Confirm that your presentation materials are good to go. Assess the setup of the room. Is there a microphone? Where will you stand in relation to the screen or podium? Will

---

Conversation Starters

Do you have trouble starting conversations with strangers before you give a talk? Many of us do. Here are some opening lines that Roger uses to break the ice:

"Hi, I'm today's speaker."

"What brought you to the meeting today?"

"How long have you worked in this field?"

"What kind of work do you do?"

When you have a friendly smile on your face, even the most introverted stranger or important VIP will be glad to talk to you if you use these starters.

---

there be a monitor so you can see your slides in front of you? Get situated, *but don't sit down.*

Get Acquainted

Now is your chance to get acquainted with people in the audience. Walk around, say hello, and introduce yourself as the speaker. Smile, extend your hand, and strike up a conversation that is relevant to the topic of your presentation.

Chat Purposefully

Ask a question that elicits a common-ground answer that you can use. "What made you decide to attend my talk today?" is an easy opener. Show appreciation for the chat and thank the person by name. Meet as many people as time permits. Now that you have broken the ice, you won't feel like you are talking to a group of strangers.

Amy's colleague Jeff Johnson has warming up the room wired. He is a retired fire chief and former chair of Oregon's Statewide Interoperability Executive Council and has extensive public safety experience at both the local and national level. Chief Johnson played a leadership role in the campaign to convince Congress to set aside radio spectrum for the first nationwide public safety wireless network. The need for a dedicated broadband network for this purpose was the last standing recommendation of the 9/11 Commission. Thanks to a united effort involving every facet of public safety, The Middle Class Tax Relief and Job Creation Act of 2012 established the First Responder Network Authority (FirstNet) as an independent authority to serve emergency responders. Chief Johnson was named to the FirstNet board in 2012 and became vice chairman in December of 2014. Amy had the pleasure of working with him in the early days as a consultant helping to tell the FirstNet story.

What inspired Jeff to champion FirstNet? In the aftermath of the 9/11 attacks, 343 firefighters lost their lives. Network interoperability, multiple technologies, and multiple proprietary approaches highlighted the dysfunction in public safety communications during the crisis. Communication network shortcomings continue to cost firefighters their lives. On June 30, 2013, nineteen members of Arizona's Granite Mountain Hotshots crew of specialty firefighters died while trying to control a blaze near Yarnell, Arizona—the highest death toll for firefighters battling a wildfire in the United States since 1933. It was the worst loss of life in a single day for U.S. firefighters since 9/11.

Only the lookout survived. The other nineteen Hotshots died when they were overcome by a blaze that suddenly changed direction and grew in intensity.[1] A report by the Arizona state forestry division found that a communications blackout of more than thirty minutes just before the men died meant their colleagues were unaware of the team's location and intentions.[2] Jeff Johnson looks forward to the day when a drone streaming a live thermal image, coupled with wind, weather, and topography data, will make it possible to prepare real-time projections of where a fire is going to go. That's what a robust data network will do for an incident commander. He explains the vision for FirstNet:

> It will be the backbone for vital innovations. We will have real-time situational awareness of our battlefield for incident commanders. They will have a common operating picture integrated with important information such as weather, topography, and transportation routes. FirstNet will make it possible to see where emergency responders are located. You could have a voice communications blackout like the situation in Yarnell, and FirstNet will provide visual redundancy. FirstNet will augment existing voice communications and provide a reliable broadband data network that is mission critical.[3]

Despite the undeniable need, FirstNet had its share of controversy in 2013. It stemmed in part from concerns over whether wireless industry veterans would truly look out for the needs of firefighters and law enforcement or just seize the opportunity to line the pockets of the major wireless carriers. An early report from the board's chief technology officer had a title that made it sound like it was the preliminary network design, which led public safety members of the FirstNet board to falsely assume that the network design had already been determined, even though FirstNet had not yet met its obligation to consult with states about their public safety needs. While it wasn't required, FirstNet also chose to include the 566 federally recognized Native American communities in the consultation effort since network design would cover large portions of their lands. With so many players in the mix and an opt-out provision that said states didn't have to use FirstNet, championing this new project was fraught with communication complexities.

It was clear that FirstNet needed champions. The organization had to bring people together, explain its vision, outline the process, and start building trust and support early on. With that goal in mind, First-Net held meetings around the country to get connected to its constituents. Jeff was the keynote speaker at a FirstNet-sponsored meeting in Denver of one hundred public safety representatives from ten states. Amy was part of the team that developed the talking points and presentation materials and rehearsed the executives before the meeting.

Jeff's mission was to discuss the approach for planning FirstNet and allay concerns at the state level. To build trust, he needed to set the tone for frank talk and transparency. He also wanted to give credit to members of the public safety community for their tireless efforts to convince Congress to allocate radio-frequency spectrum and funding to build FirstNet. He knew he needed to warm up the room.

One of the people Chief Johnson talked with on that May day in Denver was a reporter. Paul Kirby from Telecommunications Reports Daily (TR Daily) was there to cover the first part of the meeting, which was open to the press. Paul knew that when the interactive workshop portion of the meeting started, reporters would be asked to leave. As he opened the meeting, Jeff created some levity by joking that Paul had said he was planning to impersonate a FirstNet board member so he could stay. Jeff also acknowledged some very important people in attendance who were part of the fight on Capitol Hill to secure spectrum for FirstNet. He used this recognition as a lead-in to his remarks:

Good morning. I was talking to Paul Kirby of TR Daily earlier, and Paul said, "Hey I know I'm supposed to leave at 10:30 a.m., but I might try to sneak back in looking like Chuck Dowd." I said, "Hey

Paul, I just got a text from Chuck Dowd, and he's going to be here today, so if I see two Chuck Dowds in the room, you're busted, bro."

Also, I want to recognize a couple people that made substantial contributions to us being here today—Jimmy Jinardo and Chief Charles Warner—two people who contributed mightily to this workshop today. Both sit on the Public Safety Advisory Committee. I also want to acknowledge Ken Bowley. Ken is not a guy that seeks the limelight. The fact is when we were on Capitol Hill, Ken gave us excellent strategic advice at important times. And his background in terms of contributing to build out in the DC area, his experience, and even sometimes his lawyer skills, were very helpful.[4]

With these remarks, Chief Johnson ensured that important members of the audience knew he was taking them seriously. And he knew who was there because he had carefully canvased the room before he began his talk.

PUT PURPOSEFUL CHATTING TO USE

How do you use what you learned while purposefully chatting? As Jeff showed us, it can work well in your opening remarks. You might say something like, "I was talking with Kevin Saunders about the experiments he's been running, and I was pleased to hear that his results were similar to ours." Reference something relevant to your topic. Tie it together. Mentioning Kevin by name gets everyone's attention. Others will wonder if you might mention them. You've already made a very important connection with your audience. You could also weave the connection to Kevin further into your talk as you describe your experiments instead of leading with the reference.

Caution. When you cite someone's comments, do so factually. Report only what you heard. Don't embellish or interpret, or you run the risk of getting off to the wrong start. "Similar results" is a safe characterization since it is broad, but you might be tempted to be more specific and call out a more detailed finding. You could do so if, and only if, you had enough time to understand the results and how the experiment was run.

If you have partners or collaborators from a team in the room, it's always a win if you acknowledge them, just as Jeff did with the people on the Public Safety Advisory Committee and those who helped convince Congress to create FirstNet. Sharing the limelight sends a subtle message of self-confidence and appreciation. You can mention everyone by name if the list is short and you are not likely to forget anyone. The other way to acknowledge these individuals is to ask them to stand. That way, people can seek them out after the talk. You can do both if time permits. Just don't wade through a long list of names in your opening remarks. Save that for later. Use your opening to make listeners care about your topic.

Warming up the room may also help you discover how to modify your talk. If you find that you have a technically sophisticated audience, you may want to go a bit deeper into your subject matter. If other scientists in attendance also work in your field, you will want to be careful to sound competent but not come across as a know-it-all. You might also realize there is an additional topic you want to touch on based on something you learn from your warm-up conversations.

Do your best to find out if there will be any decision makers in the room before you arrive. If so, find out who they are and try to arrange an introduction. An excellent question for them is, "What part of this topic are you most interested in?" Even if they come back with an answer that is not covered in your slide deck, you can acknowledge their interest at some point in the talk to show that you are paying attention to what matters to them. There is also value in acknowledging any other scientists in your field that are in the room. This is not only a generous gesture; it also helps build relationships and may lead to meaningful collaboration.

Sometimes, warming up the room yields a clever or humorous start to your conversation, like Jeff's banter with the reporter in the audience. Steve Bohlen had a similar experience. Steve has served science and society as a prominent researcher, professor, senior manager in both federal and California state government, CEO of a systems engineering and naval architecture firm, and a scientist at Lawrence Livermore National Laboratory, where he leads programs that advance energy technologies and energy security. Steve recalled the time he spoke to the Fixed Income Forum about hydraulic fracturing. As he was visiting with people before his talk, he identified an unexpected connection with a gentleman there and used it to inject a bit of humor in his opening and engage his audience: "I did notice that the head of TIAA [the financial services company] is here. Half of my retirement is with you, so please pay attention."[5] We will hear more from Steve later in the book.

INTROVERT OR EXTROVERT: ADJUST YOUR APPROACH ACCORDINGLY

Knowing whether someone is an introvert or extrovert can help you effectively champion your science and make connections. In general, if a person is extroverted, you will likely find they think out loud and are comfortable asking questions. They typically prefer dialogue and conversation about a topic. Introverts, however, tend to want more time to process new information, so be sure to pace yourself accordingly. It may be wise to send background material in advance so they have time to think about it and feel more prepared to have a discussion with you. Introverts think things through before speaking, which can be misinterpreted as disinterest, especially by extroverts. Typically, introverts don't enjoy small talk and much prefer in-depth conversation on a topic of substance. They tend not to be big risk-takers, so keep that in mind as you think through your approach. They also prefer one-on-one discussions, so leave your entourage behind if you are meeting with an introvert.

If you are an extrovert talking with a slow-speaking, quiet introvert, you would be wise to meter your pace, pause between ideas, and turn your volume down so you don't overwhelm your listener. The reverse is true if you are an introvert trying to interest an extrovert. Mirroring communication styles paves the way for building rapport, trust, and a willingness to listen.

Obviously, no one puts "introvert" on their LinkedIn profile. But their friends and colleagues usually know, so you might ask someone you both know. Author Susan Cain talks about introverts—like herself—in *Quiet: The Power of Introverts in a World That Can't Stop Talking*.[6] She gives a quick quiz based on the core characteristics of introverts in an interview with NPR that you can find online.[7] When dealing with decision makers, it is always acceptable to ask their staff or deputy how they like to receive information. Just like looking up their resume, asking questions like this gets you major points because it shows that you are using the decision maker's time effectively.

DIFFERENCES IN LEARNING STYLES

Julio Friedmann reminds us to consider the diversity of learning styles: auditory, visual, and kinesthetic. When he was a deputy assistant secretary at the Department of Energy, he dealt with a myriad of decision makers. In our interview with him, he explained,

> Some people get information best with pictures and some people get it in writing. If you're going to give a fairly information-rich presentation you should communicate with all these things—your spoken words, text, and images. That way most people in the audience have a chance to get it on their terms. And normally with

text on slides, if you can't say the item with eight words or less then don't bother, because a paragraph is too long to absorb quickly. There was a company I was working with, whose CEO got information by reading. If he didn't read the words, he wouldn't get what you're talking about. I [handled] that two ways. Before I gave a presentation I wrote something up for him. My contact said, "The CEO's limit is a two-page white paper, it can't be three pages. Make it two and he will read it." The second thing I did was make sure that there was text in the slides so that he would read the main ideas. Everyone has different degrees of tolerance for spoken and written words in a presentation—try to find out ahead of time what is best for your listeners. And respect the fact that there is always a mixture of learning styles in every audience.[8]

Also remember that no one ever wants to look stupid. Your success as a champion depends on your ability to share information on a level that makes sense and balances your expertise with the needs of your listeners. It is also helpful to flex your communication style to make your listeners more comfortable.

TALKING TO POLITICIANS

A former senior congressional staffer[9] told us that members of the U.S. Congress love to talk to scientists because they like to understand new things. Many elected officials are curious and bright people who appreciate learning new information. But they find it frustrating when scientists communicate poorly. His advice was to never use acronyms—even if you are talking about the United States of America, don't say U.S.A. Having spent a lot of time listening to scientists brief congressional representatives, he also said it is very important to take it slow—explain things clearly in common language, and do not ever say, "You are familiar with this concept, right?" The member will nod and say, "Of course," but that is because they never want to admit that they don't know something. Consequently, the rest of the conversation may be lost.

The staffer said that he lost count of the number of times that a scientist used too much jargon or cited some technical paper ("This, of course, is based on the groundbreaking work of Smith and Jones, 1995") when briefing a representative. The member made affirmative responses—"Yes, I understand. Yes, I see your point"—but once the scientist left the room, they would turn to the staffer and say, "Well, that was a waste of my time."

Confirming and politely checking in on understanding can be difficult. It is generally better to try to evaluate what the member knows ahead of time and make sure to start from that point. If you aren't sure how much a politician knows about a technical area, assume it is not much. Ask the staff. They want their boss's time to be used well, and they will help you. If your organization has a government relations liaison, check all your material with them.

Julio reminds us to explicitly consider the job of the person you are briefing:

> There's a different approach for briefing a senator . . . [than] one of their staffers. The staffer's job is to tell the senator what he or she needs to know so they can make a decision. Staffers are likely to take more time with you and ask a lot of questions. You can bring a lot of information, but you've got to be able to drill it into their heads within 20 minutes. You need to figure out how to give them information so that they can give it back to you, because if they can give information back to you, then they can convey it accurately to the senator.
>
> Senators have a different job. They have a lot less time. You get maybe ten minutes if you are lucky. Senators will want to know why am I talking to you? What information do I need to know? Am I going to need to make a decision? What action do I take? The action might be to book a committee hearing. It might be to start working on draft legislation. They just want to know what decision to make based on the information you have provided. It's about meeting their needs; it's not about the science.

Steve Bohlen adds some very important words of wisdom for addressing savvy members of Congress:

> You have to do your homework and find reliable research reports that back you up. Be conservative. Never oversell. The one thing you never ever do is try to bullshit a member of Congress. Never use information that isn't verifiable, 100% backed up and solid. Members of Congress have good bullshit detectors and they are likely to turn around and say they would like to know more.
>
> The only thing you have is your credibility. If you are called on it, you have to be able to produce. Don't inflate numbers. Don't use something that you can't back up. Ask yourself, if this story appeared on the front page of the *Washington Post,* would you stand by it?

TALKING TO VENTURE CAPITALISTS

You may make connections with venture capitalists at conferences or events, but the conversation takes on a new level of importance when you've impressed them with your business plan and they want to meet. A business plan is the key starting point for serious discussions with a venture capital firm—they invest in businesses, not science. But we will leave discussions of business to other sources and focus on how to communicate the science in your business.

We interviewed Nancy Floyd, founder and managing director of Nth Power, the pioneering energy venture capital firm with investments in more than sixty companies, including some of the market leaders in renewable energy, energy efficiency, smart grids, clean transportation, and green buildings.[10] A highly respected entrepreneur, Nancy has founded and helped launch seventeen companies, delivered an address on green technology to the 2008 Democratic National Convention, and served as a judge for idea competitions such as the MIT Energy Prize and many clean tech and National Renewable Energy Lab events. Nancy advises entrepreneurial scientists to do their homework and make sure they are targeting the right type of investment firm:

> In the Internet age, there is no excuse for not having some level of understanding of what the VC firm does. It's usually easy to find out what a firm invests in because most firms will publish the list of their investments on their websites. Most firms will list the criteria they are looking for in a potential investment. You don't really want to go to a firm that's invested in one of your close competitors and you look naïve at best, and unprepared at worst, if you go to a firm that only invests in companies that have revenue when you have none.

She indicates that your company's stage of development and the amount of funding you are seeking are key considerations in targeting a VC firm:

> If you think you need small dollars then you need to go to a different type of VC fund than if you need multiple millions of dollars. There's a broad range of firms. Some invest in profitable companies when there is less risk and some like to invest in pre-revenue companies. So you need to do your homework and identify which firms are willing to make seed or Series A investments.

Nancy recommends that before a first meeting, scientists find out what the people at the firm know and what they are like as investors.

She suggests working through the firm's partner that has been your primary contact and being sure to ask them what you should cover:

> Know your audience because that's going to drive the type of presentation you give. In a first meeting, you've only got 45 minutes to cover the technology, market opportunity, the customer, competition, sales strategy, who you need to hire and how much more you'll likely need to raise until you are mature enough to get bought or go public.

Nancy cautions that time management is crucial:

> If you spend 45 minutes or even 30 minutes covering material that—if you had done your homework—you would know that the investors already understand, then you've just wasted time when you could have gotten into a really good Q&A with them.

Lastly, she advises that you assess the potential funder for appropriate fit, just as they are evaluating you:

> Know your audience as potential partners, too. You will be working with anyone who invests in your company so it's wise to do your own due diligence before you are joined at the hip. I've seen horrible things in the boardroom—like oil and water between the management team and the investor.
>
> You are basically sharing your crown jewels with somebody. So I want to encourage scientists and founders, as you get into the process with a potential investor, to figure out if the investor is really somebody you can work with, just like we are figuring that out for ourselves.

INFORMAL AND UNEXPECTED AUDIENCES

Amy often reminds her clients that every time you communicate, you have an opportunity to be strategic. The key is to think that way and be mindful of the message you want to deliver to advance your science. Roger recalls a missed opportunity that highlights an important principle of championing science: share what you can about your work every chance you get because you never know who is listening and how much they might help you.

Roger was at a Department of Energy (DOE) meeting, seated with eight strangers enjoying rubber hotel chicken. The lunch conversation about new types of environmental and energy technology was brisk. After discussing a carbon capture demonstration going on in North Carolina, he turned to the dapper older gentleman next to him and introduced himself. His neighbor returned the pleasantry, telling Roger he was working at a major Southern university after having spent several decades at a flagship industrial firm. He was at the meeting to try to get a sense of whether the DOE might be interested in his new technology, which converts carbon dioxide (and apparently other oxides) into useful compounds.

At this point, he had captured Roger's interest. The gentleman had had a long career, which could be expected to provide a lot of useful experience, and an application (converting waste carbon dioxide into useful products) that intrigued Roger. Anticipating a lively and informative discussion, he asked, "So, how do you do that?" The man's answer: "I'm sorry, I can't tell you. We are here to talk to the DOE folks, and we haven't received all the patents yet."

Roger hears this all too often. There's a magic technology that is going to save the world and make the inventors rich. But he was surprised to hear it from an experienced and worldly looking academic. Roger pressed on. "What products do you make?" was also rebuffed. The man wouldn't even explain what the process *did,* let alone what it *was.*

In this case, Roger's lunch partner missed several valuable opportunities. First, he missed the chance to try out his pitch. The more you test your pitch on new listeners, the better it gets. There were several other experienced players in his field at the table who would have made a great impromptu practice audience.

Second, the gentleman failed to recognize that the pitch you make directly to a funding agency is never as powerful as the pitch *someone else makes on your behalf.* When experienced people hear about interesting new approaches, they are delighted to mention them to program managers—who are often friends and value their input. Of course, this gentleman had no idea that Roger could play that kind of direct advocacy role, but that misses the point. In our connected scientific world, *anyone can have that kind of influence for you.* At a focused technical meeting, the likelihood that an obviously senior (well, grey-haired, anyway) person like Roger could play that role is very high.

The more people who know about your idea, the better your chances of building support. *Secrecy is rarely as valuable as collaborators and supporters.* Tell your story to everyone who will listen, and pay attention to their feedback. Seek out potential influencers. Become a better champion and bring your game-changing idea to fruition.

5

Why Scientists Communicate Poorly outside Their Field

The main cause of incomprehensible prose is the difficulty of imagining what it's like for someone else not to know something that you know.

—Steven Pinker, *The Sense of Style*

If you want to learn to communicate effectively, it is useful to understand *why* scientists find it hard to talk to the public and especially to decision makers. We are all good at talking to our colleagues and professors and progressively more dismal at communicating to people with less and less of our training. And it is not because those people are stupid. That lame explanation is brought up too often when scientists talk to each other about this problem. It is neither true nor helpful. What is true is that magnificently intelligent people may know little about science. That doesn't mean they don't have the capacity to understand. It is incumbent on us to make science understandable.

Think about your most recent presentations. Chances are they were delivered to other scientists in your field and you gave only the briefest introduction of why the material was important. It's also likely that your introduction was couched in terms accessible only to other experts in your field—"The need for a better equation describing this key relationship is well known to all of you." Scientists are trained to be exceedingly good at detail and rewarded for this as students. Meanwhile, they usually assume that the big picture is understood. This is a major part of why scientists have so much trouble communicating with decision makers.

But there are factors that are more deeply embedded in scientists' training that get in the way of communicating well. Two of the most important are intrinsic parts of communicating about any richly endowed topic: *jargon* and *paradigm*. Jargon has to do with words, and paradigm has to do with ideas.

JARGON: THE SPECIALIZED WORDS OF SCIENCE

Jargon enables rapid, accurate conversations about any specialized topic. In fact, we use jargon in our everyday lives, but since we share that jargon with others around us (at least regionally or culturally), we simply call it language. One of our favorite users of jargon is the armed forces, where precise, long phrases like "Commander in Chief, Pacific," are transformed into short acronyms like CINCPAC for everyday military usage. Eventually, these condensations are no longer seen as acronyms but as new nouns, accessible only to those with special indoctrination. Entire paragraphs can be condensed into a sentence by this method. If you have talked to anyone in the military, you will recognize how useful it is for rapidly discussing complicated organizational structures. The obvious downside of such a powerful communication method is that it requires the listener to be fully trained.

It is all too easy to complain about the jargon used in any field outside your own. As scientists, it is important to acknowledge jargon's importance in our work. When a doctor talks about the condition of your spinal cord, most of us would prefer hearing familiar phrases like *bony growths* or *bone spurs* rather than the more precise phrasing *osteophytes*. However, when your doctor is talking to the neurosurgeon about exactly where they are going to cut, it is imperative that they use an extremely precise set of words to describe the problem and what they are going to do about it. Most physicians would avoid using full jargon, without any translation, in front of a patient. Medicine has acknowledged the difference between these two classes of communication. They recognize that in both cases, failure to communicate using the correct level of language can lead to bad outcomes.

This issue of naming things accurately is at the heart of science. Prior to the mid-1700s, plants were given complex Latin names that were descriptive but did not imply relationships among species. Often, a single plant had multiple names. Carl Linnaeus created a naming system that was both specific and relational—similar species were included in orders (e.g., humans are part of the order Primates), which implied similarities in morphology, structure, and, as was later discovered, evolution. Linnaeus's hierarchical classification and binomial nomenclature, although much modified, have remained standard for over two hundred years.[1]

This "naming phase" is crucial to the success of any multidisciplinary group because, as Linnaeus showed, you can't talk about important items until you can name them. You will quickly discover that one of the first important tasks for a new team is to agree on how to talk about the subject. What do specific words mean, particularly those that have slightly different meanings in different disciplines? Once your team agrees on nomenclature, you will notice a dramatic increase in productivity, and you will have a decidedly smug feeling about your team's newfound ability to collaborate among fields where others can't even talk to each other.

Learning names is one of the first hurdles in being educated in a scientific specialty. Early on, learning all the specialized terms seems insurmountable, but with time those terms become more than random syllables. We begin to associate complex phenomena with them, and like the soldier shortening his chain of command, we scientists begin to think of those complex phenomena *only* in terms of the special terms—jargon—that we have learned. Even if a simpler term would suffice, we tend to use the precise language that is not only brief but also conveys a deep set of meanings. Many of us have never spoken in general terms about broad aspects of our work. We only know it in terms of our special language and have had little occasion to try out how that language is understood by those outside our field.

Medical doctors talk to nonspecialists every day, so they have ample opportunity to gauge the effect of their communication style. But what about the rest of us, who don't have a regular obligation to test our ability to be understood? How can scientists learn to use the correct level of specificity? Once again, this requires being self-aware. Does your audience share your understanding of jargon and special usage? It is safe to assume that decision makers rarely understand much of your discipline's unique language. It is magical to be able to speak quickly with fellow wizards, but you have to absolutely avoid jargon when championing science. How can you talk about difficult subjects without using the precise terminology you have learned? Here are some specific suggestions.

Spell It Out

Use the full phrase, not the acronym. If it pains you to type them out, set your word processor to automatically expand your common acronyms. If you feel compelled to use an acronym during a talk, do so, but immediately spell it out and say it in full every time you mention it. Pay attention to the terminology that your audience knows.

Most importantly, never rely on acronyms that are unfamiliar to your readers to shorten your text to meet a word limit. If a reader has to struggle to understand your language, they have less brainpower available to understand your message or proposal. Enormous levels of detail will not help if your basic message is lost.

Convey the Full Richness

Remember that much jargon is not just a condensation of multiple words. There is often a mechanism, system linkage, or evolutionary foundation (as was the case with Linnaeus's classification scheme) at the heart of the jargon. As a geologist, when Roger talks about mid-ocean ridge basalt (please not MORB!), he does so with an understanding of the full context of everything he knows about plate tectonics and the formation of the current surface of the planet. All of that is wrapped into four words, *mid-ocean ridge basalt,* which are fully understood by any-one in his field but almost entirely lost on anyone who thinks of rocks as something that you find in a creek.

If you mean to convey that kind of richness, take a moment and praise the way that "volcanoes that we'll never see build the ocean depths, the very skin of the earth yanked apart to release the molten lava beneath, the continuous tearing driven relentlessly by the ultimate death of the opposite side of the plate as the cold, old surface of the earth is dragged down once again into the furnace for a cycle of rebirth." Perhaps it will sound theatrical to your peers, and they may kid you about it later, but some will be sorry they didn't take this approach themselves. More substantively, your lay audience will be left with a strong sense of how important those rocks must be.

Don't Overload Your Listeners

Don't define a special term or acronym, like *spinal stenosis,* and then blithely continue using it in the rest of your discussion, assuming that nonexperts have instantly learned it. You can generally expect most audiences to learn *one* new term, but even then, you have to take a moment to clearly describe it. Tell the audience why the term is impor-tant to the rest of your discussion. Highlight that it is something you expect them to learn. Emphasizing it a bit in your presentation materi-als, such as by italicizing *spinal stenosis,* lets the audience know that this is a special term to which you attach heightened importance.

Use Mixed Phrasing for Mixed Audiences

In mixed audiences, you may feel that using a common, explanatory phrasing will cause experts in attendance to think you are a nonexpert. We recommend that you pair specific and nonspecific phrasings. For example, "We are studying the effects of spinal stenosis, the narrowing of the bone channel where the spinal cord runs." This technique will keep both the expert and nonexpert fully informed. It is an artful way to be respectful of all the members of your audience while keeping your reputation intact in the presence of any science snobs. This is not dumbing it down.

PARADIGM: THE ORGANIZATION OF SCIENTIFIC THOUGHT

The words that scientists use are just a superficial representation of the detail required for scientific communication. We touched on this complexity in the discussion of mid-ocean ridge basalts, but the linkages extend to the very edges of disciplinary knowledge. This network of understanding can be called your *paradigm.*

A paradigm is the shared experience of those in your field. It is perhaps more insidious than jargon because scientists fail to recognize how shaped they are by the extensive learning and discovery in their specialized area of study.

While the word *paradigm* has evolved many meanings in popular culture, the original family of meanings we use comes from Thomas Kuhn's timeless work *The Structure of Scientific Revolutions.*[2] While a work like *Structure* may seem far afield from championing science, Kuhn lays out a cogent explanation of the internal workings of scientific communication within groups of like-minded scientific practitioners. This is an excellent way to understand the *lack* of communication outside these groups or, more broadly, outside these paradigms. According to Kuhn, "A paradigm is what the members of a scientific community share, and, conversely, a scientific community consists of men who share a paradigm. . . A scientific community consists, on this view, of the practitioners of a scientific specialty. To an extent unparalleled in most other fields, they have undergone similar educations and professional initiations; in the process they have absorbed the same technical literature and drawn many of the same lessons from it."[3]

Ian Hacking expanded on this in an introductory essay to Kuhn's book: "You are inducted (into your paradigm) not by the laws and the theories but by the problems at the ends of the chapters. You have to learn that a group of these problems, seemingly disparate, can be solved by using similar techniques."[4]

After examining a large number of scientific revolutions, Kuhn argued that perception *cannot be separated* from scientific reality. A scientist sees science only in the context of his or her paradigm. Listeners from outside the field have a very similar control of their own perception, shaped by their own paradigms. Not only do these scientists not share the same words with those who don't share their paradigm, but they also have *fundamentally different views of the world*. Kuhn gives an example of a map: a student sees only lines on paper, but a cartographer sees a picture of the terrain.

It is an aside to our efforts to describe communication, but Kuhn's fascinating realization was that when a scientific revolution occurred—say when Lavoisier heated the red oxide of mercury and recognized that the oxygen he obtained was a gas that was distinct from air—practitioners of the new paradigm suddenly see old data in new ways. In Kuhn's words:

> Lavoisier. . . saw oxygen where Priestley had seen dephlogisticated air and where others had seen nothing at all. In learning to see oxygen, however, Lavoisier also had to change his view of many other more familiar substances. He had, for example, to see a compound ore where Priestley and his contemporaries had seen an elementary earth, and there were other such changes besides. At the very least, as a result of discovering oxygen, Lavoisier saw nature differently. And in the absence of some recourse to that hypothetical fixed nature that he "saw differently," the principle of economy will urge us to say that after discovering oxygen Lavoisier worked in a different world.[5]

Phenomena that had previously been dismissed as poor experiments or presumed to be muddled by large errors are seen to be precisely explained by the new paradigm. Even phenomena that were patently at odds with the old paradigm are recognized as having indicated all along that the old paradigm was wrong. Why were the results originally missed? Because in the old paradigm, there was no explanation for them, and the ability to perceive them was severely limited by the structure of the scientific knowledge in the brains of the practitioners. And that is the nature of truly notable scientific revolutions: when a new cascade of understanding is suddenly triggered, an entire discipline changes its perception of itself in a short period of time.

Other People's Paradigms

Scientists have evolved detailed paradigms to systematize their understanding, but most people do not have *any* paradigm for interpreting science, which makes communication difficult. Moreover, it makes it impossible for scientists to communicate with nonscientists at anything like the speed they are accustomed to when dealing with specialists within their own field. That's like forcing someone to drink from the proverbial fire hose. They can't possibly absorb everything. They are not stupid; they are uninitiated. *Failing to recognize when others do not share your paradigm is the most important error to avoid in championing science.*

Paradigms can hinder the ability to communicate by changing the listener's understanding relative to what the speaker intended. For instance, paradigms embodied in religious belief or ethical standards of behavior, which by their nature attempt to be all-inclusive, shape perception in ways that scientists dismiss at their peril. A deeply ethical person may be loathe to pursue a direction because they suspect it will conflict with their principles down the road, even though they may not be able to immediately name or evaluate that conflict. If scientists do not consider this paradigm-based behavior, they can completely fail to communicate.

Ethical Paradigms

It can be particularly difficult if the listener is technical in some sense other than the topic at hand but chooses to apply an ethically derived general paradigm when they are not fully initiated into your technical paradigm. This *gap* in understanding frustrates many well-meaning interactions, such as those between scientists and politicians or nongovernmental agencies, where the existence of deeply held, ethically based beliefs (like biosafety or environmental protection) are at the core of the listener's mental view. Those foundational ideologies are often associated with the reason they want to talk to the scientist. If you recognize this and try to acknowledge and openly address your listener's paradigm in your description, they will understand you more rapidly and completely. They may still not agree with you, but at least it will not be because they missed your message.

Compensating for Paradigm Gaps

When you and your listener have different paradigms, there is a gap between you. To address it, you need to be self-aware and self-correcting. Successful communicators think about their paradigms on three levels: 1) scientists working in their own discipline, 2) scientists outside their field, and 3) nontechnical decision makers. If you can gain an understanding of your audience ahead of time, you can probably assess which of these classes your listeners belong in and adjust your level of discussion accordingly.

The onus lies on the champion to *begin with* the language and paradigm of the listener, since it is more likely that the scientist speaks the listener's language than vice versa. This is by no means dumbing it down but rather meeting the listener where they are and bringing them along with you to experience the journey that you have already undertaken.

A prime instance of an issue that is affected by ethically based paradigms is the genetic modification of organisms. One of the earliest applications of recombinant DNA was for producing insulin. Bill Young, former Chief Operating Officer of Genentech, told us the story of his early days in biotech,[6] when Genentech had successfully developed a process for Eli Lilly that replaced pig pancreas extracts. At that time, there was no general understanding of the risks of culturing *E. coli* with altered genes that expressed human insulin.

Bill noted that people eventually thought, "Okay, if you make these organisms capable of making a protein, you might create something dangerous. What if they get out into the environment and cause unintended problems?" The U.S. Food and Drug Administration set up a committee at the National Institutes of Health to review and approve experiments with recombinant technology. "If you wanted to produce more than 10 liters," Bill explained, "you needed special approval. You had to get a recombinant DNA Advisory Committee approval to exceed that scale. And of course, anything we were doing on the process side had to exceed that scale." He remembers going to those committee meetings and showing how the fermenters were specially designed to make sure that nothing could escape. It took a whole round of simplified communications and visuals to allay a lot of fear and ultimately enable the experiments to go forward. That went on for a number of years. Ultimately, nothing bad happened, and the guidelines were eventually lifted.

By 2017, many drugs, including new cancer therapies and hormone replacements, were being created through a process of *E. coli* expressing some other species' genes to create a useful protein, but in the 1980s, there was no shared paradigm with which to understand the risks and benefits of the approach. The presence of unknown risks was a major impediment to commercial deployment. In the absence of the shared paradigm that regulators and drug companies have today, an ethical paradigm—the *precautionary principle*—was predominant in the conversation. Science champions are sure to encounter this situation. The precautionary principle is a common paradigm, and much experience has shown that it can be valuable in avoiding disaster.

In cases like this, the audience has a preformed view of the world, and it is the champion's job to supply the information and understanding necessary to reshape this paradigm. The more extreme these gaps between paradigms are, the longer it takes to create shared understanding.

INFORMATION DEFICIT THINKING AND
THE OPINIONATED DECISION MAKER

Scientists tend to believe that giving someone more information is likely to bring them around to their way of thinking. This is broadly known as the information deficit model. Bill Young had a relatively easy job convincing the U.S. Food and Drug Administration; a mere decade of meetings, experiments, papers, and outcomes was all that was required. This is the classic approach to information deficit: the scientists bring the facts, and the world accepts their assessment.

Today, climate change is the primary topic where scientists would like to believe that more information will sway more minds. Nuclear power and vaccines are other examples. But "alternative facts" are increasingly popping up in many venues, and it is clear that not all opinions can be overridden by the mere application of truth.

This book is intended to instruct scientists on how to communicate with informed decision makers, not the general public. Still, the problem of entrenched ideas, like ethical concerns, is an issue when talking with anyone. However, unlike dealing with the problem of paradigm offset, there are many situations that call for a more thoughtful approach than just piling on more and more information.

Why doesn't it work to simply inform people? It turns out that a principal problem is that there are too many experts. Dan Kahan, a psychologist at Yale University, has tried to understand why opinions about climate change have remained so rigid in the United States despite a decade of scientific information.[7] Kahan writes that when people form opinions about climate change, "They selectively credit or discredit evidence in patterns that reflect their commitments to important or self-defining social groups."[8] In other words, they tend to go along with the prevailing opinion among their friends and peers *because to do otherwise would threaten their place in that social grouping.* And so they grasp at any professed opinion that matches their preconception and assign it the weight of an expert opinion. It's not that they eschew expert opinions, they just redefine them in terms of their existing bias.

Katharine Hayhoe is a climate scientist at Texas Tech University. She is on the front line of the climate debate in the United States and has developed some of the most compelling material discussing what climate change is and what we can do about it. She has created a series of short explanatory videos called Global Wierding. In "If I just Explain the Facts They'll Get It, Right?"[9] Katherine confronts, head-on, the issue of how to talk to people when the reason they don't believe climate change is real is based on identity and ideology rather than facts. In this case, arguing over facts can seem like a personal attack.

Katharine's strategy is not to bombard with data and facts, but to first bond over shared values, such as children, business, or hobbies. Then, connect those values to climate change. For instance, in a discussion with a business-oriented person, she would mention that in Texas, a solar or wind power plant provides eight times more jobs than a natural gas plant. This also highlights the benefit of offering a solution. When confronted with a difficult problem that doesn't seem to have a solution, people tend to feel disenfranchised and powerless. They disassociate from the problem and deny it. Suggesting a practical solution brings them back into the discussion.

Katharine offers a great iconic analogy as well. She explains that climate models are based on the same physics as refrigerators and airplanes. If we deny the science of climate change, are we going to do without cold food and air travel because we don't believe in the science?

Brendan Nyhan, a political scientist at Dartmouth, has looked at this problem of entrenched ideas in the broader context of general political discourse. What mechanisms would work to avoid this ingrained problem? In a series of studies, he examined the effectiveness of two approaches: transmitting information in a graphical format and "reducing the identity threat posed by attitude-inconsistent facts by buttressing people's self-worth, allowing them to more thoughtfully or carefully consider the evidence."[10] The idea of using simple graphs reflects the fact that in the face of an overwhelming cultural bias against a line of reasoning, it is difficult to find the intellectual capacity to understand text or complicated descriptions.

Buttressing the self-worth of individuals proved to be of little value, but simple graphical descriptions were uniformly useful, especially when presented as a function of time. In studies of the effects of the January 2007 troop surge in Iraq, job growth during the Obama administration, and global temperatures, Nyhan and his coauthor Jason Reifler found that conventional Republican/Democratic divides on these topics could be significantly bridged by showing people simple line graphs of the phenomena in question. The use of the graphical depictions reduced the overall level of misconception (for instance, that global temperature change is not occurring) by a factor of two to three in each study. These are huge changes for a study of this kind.

Roger often lectures about the rapidly changing condition of the energy system in the United States. He has found that Nyhan's simple graph hypothesis works well. Here are three graphs that he uses to describe how utility and power executives make decisions about future investments in electric power generation. He used a third-party source for the data, the investment firm Lazard,[11] but he recast their data in a simple form, namely how much power (the product they sell) an investor could expect to get from a new power source—coal, nuclear, wind, solar, or gas. He shows three slides in sequence to expose the concepts in an understandable way and save the big surprise for the final slide.

The first slide (figure 4), showing coal and nuclear values, tends to confirm the expectations of the audience. The cost of producing these two power types has not changed much in recent years, and the 10 kWh of product per dollar invested corresponds to the price that electric power is routinely sold for. Then he adds the gas line in a second slide (figure 5), and it becomes clear that gas is a better value for investors, which is reflected in the current widespread construction of gas-fired electric generating facilities in the United States. At this point, the audience understands the graphs, and their *confirmation bias* has been engaged—they are seeing what they expected to see.

The third slide (figure 6), which adds wind and solar values, is always a surprise to the audience. It challenges their confirmation bias, as most would have said that (absent subsidies) wind and solar were less valuable to an investor than gas and coal. But having accepted the premise of the graph from the first two slides, they are much more likely to

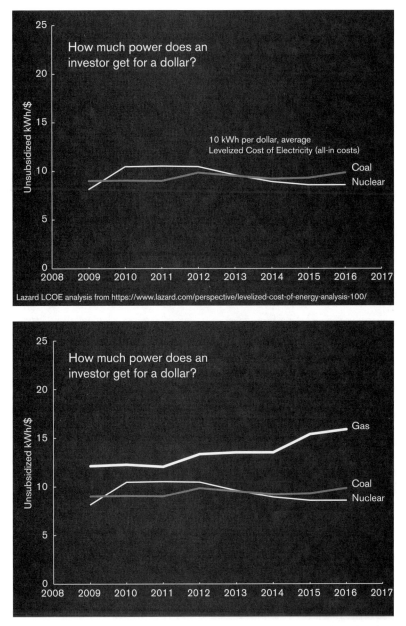

FIGURE 4. Slide showing the value of coal and nuclear electric power, based on data from the investment company Lazard. The levelized cost of electricity (LCOE) is the all-in cost, including fuel, construction, finance, maintenance, and decommissioning. Figure by Roger Aines.

FIGURE 5. Figure 4 with natural gas data added. The data source note is omitted in this second slide, and the gas line is emphasized to highlight the improvement over coal and nuclear. Figure by Roger Aines.

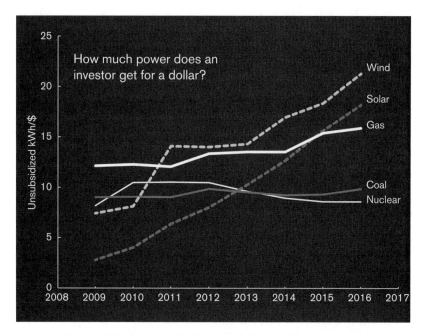

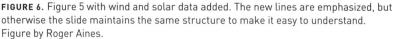

FIGURE 6. Figure 5 with wind and solar data added. The new lines are emphasized, but otherwise the slide maintains the same structure to make it easy to understand. Figure by Roger Aines.

accept the conclusion that wind and solar power are becoming the most valuable investments in electricity generation. Rather than arguing with his audience's perceptions of fossil fuels, Roger engages them and then presents the renewable picture in exactly the same terms. In chapter 9, "Designing Effective Visuals," we will return to methods for producing simple, easily understood graphics like these that engage your audience rather than challenge them.

It should not be surprising that simple explanations are the most effective—that is a theme of our book. When dealing with a decision maker who has a bias embedded in their mental process, do not try to change their entire worldview. Focus on a simple, pertinent aspect that is directly relevant to the work you need them to evaluate. Do not drown them with facts.

PART TWO

Mechanics of Championing Science

6

The First Two Minutes

Show Your Passion

Why was Carl Sagan so captivating? What makes us look forward to spending an hour listening to Neil deGrasse Tyson? Why are Jana Levin's descriptions of the universe so enthralling? Passion! The acquisition of scientific knowledge may require precision and serious discipline, but ultimately, people pursue science because they love it. Conveying your passion for your science makes you a better champion. You don't have to be overly animated or expressive; your words and tone just need to show genuine enthusiasm. We recommend that you take time to understand what sparked your own passion so you can pass on the flame.

When people understand how the work that you are doing can change the world, they get excited. Susan Weinschenk is a practical psychologist who has spent her career studying what motivates people. Her consulting business is oriented toward designing things like websites and presentations, where you essentially have one solid shot at getting a message across, either on a screen or a slide. She gives another reason for showing your excitement about the possibilities of your work: "People imitate your emotions and feel your feelings. If you're passionate about your topic, this excitement will be contagious for the audience. Don't hold back."[1] Don't be ashamed to be thrilled to be a scientist and to be doing things that really matter. Passion is the difference between boring and impressive. Excitement is born out of possibility.

Your goal in the first two minutes of a presentation or discussion is to generate contagious enthusiasm for your subject and make your listeners care. Create some attention-grabbing opening lines that will connect them to your topic. Then explain why your work is important. Help listeners appreciate the value of your science and what inspires your pursuit. Next, bring conflict into your story. In science, there is inherent conflict between the previous understanding of the world and how your work is changing it. Show your commitment to your work and outline the kind of impact you aim to make. This will motivate your listeners to care and pay attention.

The first two minutes of a discussion should frame the topic you plan to unpack. Give a sense of the destination without giving away specifics. Steve Bohlen, whom we first met in chapter 4, "Who's Listening? Know Your Audience," has been a professor and was the Associate Chief Geologist of the United States Geological Survey from 1995 to 2000. He was responsible for the research programs at that agency, which included the National Earthquake Hazards Reduction, Climate Change, Global Energy, and Minerals Resource Programs. From 2000 to 2008, he was the CEO of Joint Oceanographic Institutions, heading the National Science Foundation–funded global scientific ocean drilling efforts.

Steve says, "Give them enough to go on that they want to follow your every word. Earn their willingness to listen."[2] He offers sound advice for getting your listeners to care: "I always discuss science in terms of the value to society. There always has to be value. It's never about the science. The science is lovely to only the people who find it beautiful, but it is valuable to people and they need to understand the value."

MAKE YOUR OPENING LINE COUNT

How many conference speakers have you heard start with meaningless pleasantries? Almost all of them, unfortunately. The audience is only half-listening at this point. They know—or worse, they think they know—that they aren't going to miss anything important. We're not recommending that you don't thank your sponsor or say how great it is to be in Kansas City. Just don't use any of those things as your opening line. And please don't ever begin by reciting the names of all your coauthors— even the Academy Awards has taken to scrolling names across the screen so the presenter doesn't have to use up precious acceptance speech time.

The first words you say need to make your audience sit up and listen—even if it's just one key decision maker sitting across the desk from you. These words need to be carefully selected to accomplish

---

Engaging Opening Line Approaches

Your opening line is a powerful attention-grabbing opportunity. Don't waste it. For instance, you can spark the listener's imagination with the question, "What if you could?" This is a great way to position a long-range benefit of the work you are doing without explicitly saying that you know exactly what is required to reach that goal. You can also capture listener attention by pointing to the future or the past. For example, in a presentation about the use of artificial intelligence, you could make a prediction like, "Ten years from now, most employees will work alongside smart machines."

Revealing interesting findings, ideally new ones, works well as an opener, too. This approach gives you a chance to showcase what you have learned so far and to highlight the significance of your science. A short story about how you got involved in your work is a reliable way to illustrate challenges or state what's possible. Any approach that connects *you* to your science in a personal way turns your opening into a story that people will want to follow.

---

something vitally important to the success of your entire conversation or talk. Don't leave your opening to chance.

If you are a drug developer working on the latest medicines to fight dementia, you might start by saying, "I'm here today because my father doesn't know who I am anymore." This compelling statement does two things. It sets up the topic of your research on dementia, and it strikes an emotional chord with your listeners, engaging their hearts as well as their minds.

That's what you are after. How can you quickly build a bond with your listeners? How can your first words spotlight who you are and why you are about to take up their precious time?

Remember the story of FirstNet and the nineteen firefighters who died at Yarnell Hill in Arizona? You could start off a presentation about new emergency telecommunications technology with a dry discussion of the existing standards. Or you could face the audience and somberly say, "Nineteen dedicated men died at Yarnell Hill because they couldn't communicate." You will have the undivided attention of the audience. With that opening, your technical discussion will have real gravitas.

Here's an example of how Julio Friedmann handles the first two minutes.[3] It's a variation of the "What if you could?" approach:

Today we can provide the functions of biology, without biology. We can do what nature does in terms of energy services, like photosynthesis or making complex molecules from simple ones, and we believe we can do that without using the organisms. Imagine how that could save the world a bunch of time and money.

Julio's first sentence, "Today we can provide the functions of biology, without biology," is the kind of opening that works. That's because it's thought-provoking and makes you want to hear what's next. That's the short pitch. It's why the listener should care. Then he adds:

Humans started doing this by harnessing animals and plowing fields—the first round of energy services in nature. The second round of energy services in nature was things like the Haber-Bosch process that makes ammonia fertilizer from the nitrogen in air. It used to be that the only thing that could make fixed nitrogen was clover and legumes. Now, we have an industrial process that makes it. But it is expensive; it costs a lot of money to build the factories, and the energy conversion efficiency is awful, but you can make huge supplies of fertilizer and change the world.

Julio has succinctly set up the problem. Now he introduces the value of the science he is championing:

Now, imagine if instead of using the huge arc furnace of that process, you were able to do simple chemistry that follows the same paths as in the legumes, but with industrial scale efficiency and process density. A lot of production going on in one place, instead of spread out over a huge number of open fields. With that kind of technology you can envision a whole system where you take the core chemistry of what nature does, and repackage it. You can match nature's approach in smaller, higher output units. They can be flexible and have exactly the same environmental impact as the natural system.

Revealing interesting findings, ideally new ones, also works well as an opener. This approach gives you a chance to showcase what you have learned so far and the significance of your science. To announce the successful detection of gravity waves by the Laser Interferometer Gravity-Wave Observatory in February of 2016, the executive director David Reitze began with a simple statement: "Ladies and gentlemen, we have detected gravitational waves. We did it."[4] That was certainly a new finding!

Show-and-Tell

If you have any visual examples of your science to carry around with you, doing a quick show-and-tell bit can be a very powerful opener. It is tangible proof of your progress. Toys are fun. So are demos and prototypes.

If you have a cool thing to show your audience, pass it around *while* you describe it. In Roger's experience, people really enjoy hearing what this cool thing is *as they are looking at it*. Sure, sometimes things get lost or broken, but that is rare. As long as your materials are safe to handle, trust that people will handle them with care.

A well-told, short story is a reliable way to engage listeners as you open a talk. Here is one that explains why it's hard to predict when a breakthrough will come. In 1985, Gary Kasparov was the reigning world champion in chess. Twelve years later, he was defeated by IBM's Deep Blue computer in a well-publicized chess tournament.[5] In *The Age of Intelligent Machines,* published in 1990, Raymond Kurzweil extrapolated the performance of chess software to predict that computers would beat the best human players by the year 2000.[6] It actually happened three years earlier.

Looking into the future or the past always sparks engagement, because it taps our curiosity. Similarly, if you can provide brand new and unexpected information, you can harness the power of surprise, which, like a good plot twist, gets listeners longing to hear what happens next.

Any approach that connects you to your science in a personal way turns your opening into a story that people will want to follow. "I watched my dad battle lung cancer for eight years. Today I want to update you on my most recent research findings in the work my team and I have been doing to beat back this disease." You will find several examples in this book of strong openings with a personal medical story. That's because it works. But in any scientific field, giving the audience a sentence or two about *why* you are excited is always a good start. Decision makers in particular want to know that *you are invested in the work* and that you won't flit to some other topic while spending their money. And you don't have to reach far to find your story. You know why you love your work.

EXAMPLES OF GREAT OPENINGS: THE THREE MINUTE THESIS

The University of Queensland in Australia began a terrific competition for graduating masters and PhD students in 2008. In no more than three minutes, and using only one slide, students must describe their research in terms a layperson can understand. The format encourages the elements of extracting the essence and showing your passion, which we have discussed. Known as the Three Minute Thesis, this program has become part of most Australian universities and expanded internationally. If it exists at your school, throw your hat into the ring and try out the concepts you've been reading about here.

A useful aspect of the Australian competitions, and increasingly others, is that high-quality video recordings are made of the talks, many of which are available online.[7] All of the winners demonstrate the value of opening with passion. Some are theatrical and animated. Others are unassuming yet compelling in the words they choose and the way they deliver them. The archive of talks is huge—it is a great resource for seeing what it looks like to show passion in a style that gets your message across but still remains true to your personality.

Jenna Butler, a student at Western University in Canada, opened her Three Minute Thesis talk with an iconic analogy and a very personal reason for her interest in her science: "Picture a monster that can attack at any time, that can change itself to evade the weapons used against it, and can come back to life from the dead. It sounds like something from a horror movie, but this monster is real. It killed Steve Jobs, Patrick Swayze, and, when I was fourteen, my mother."[8] At this point, there is no doubt that everyone is listening for what she will say next about cancer.

The Three Minute Thesis is fundamentally a condensed version of the five slide presentation, covering the first three topics (problem, technical gap, and filling the gap). The winner's single slide is almost always a simple image demonstrating *why* the research is important, and contestants usually spend half of their time on this subject. Establishing the *why* is essential when introducing your science. This competition is intended to describe science to nontechnical audiences, not decision makers, but the best talks are excellent examples of how to stick to that highest level of abstraction so you can make sure the audience really understands why your science matters.

MAKING A CONFIDENT, CREDIBLE FIRST IMPRESSION

Getting your first two minutes right also requires being very conscious of how you take the stage, enter the room, or catch someone in the hallway. For casual hallway encounters, it's always a good idea to start by checking in: "Do you have a few minutes for a quick update on our latest findings?" Walk tall. Stand tall. Use a relaxed posture. Smile. Your goal is to have an inviting and confident demeanor. You also need to honor your request for "a few minutes" by finding out how much time you actually have and not going beyond it unless you are expressly asked to continue the conversation.

In most formal speaking situations, your audience will be watching as you take the stage. Walk with a lilt in your step at a pace that shows your enthusiasm, but don't appear to run to the podium. Watch the exuberance late night TV hosts like Jimmy Fallon or Stephen Colbert use when they make an entrance. Clearly, you don't want to do the over-the-top celebrity version, but it helps to see how a confident, energetic entrance is well received by the audience. Comfortable shoes that are easy to walk and stand in are essential. Smile as you approach the center of the stage and extend your hand to greet the person introducing you. A genuine smile exudes warmth, and warmth contributes to being seen as trustworthy.

If there is a podium, resist the temptation to hide behind it. You are the most interesting thing on stage. Stay in sight. Ideally you will have access to a lapel microphone so you can move around. If you have to stand near the podium to speak into the microphone, angle it to one side and stand to that side so that you can gesture freely and aren't visually cut off.

Keep in mind that you want to be near the screen that's showing your slides but not block it. Standing to the left of the screen is ideal in English-speaking forums. In some venues, you'll have a monitor so you can see your slides in front of you. If you don't have that speaker's aid, you may need to glance at the screen or angle yourself so you can gaze back at the slides occasionally. Talk *only* when you are looking out at your listeners. This is especially important during the first two minutes so you appear confident and in command of the stage. Talk to the audience, not to your slides or computer.

YOUR CLOTHES MATTER

People will judge you the minute they see you. Look the part by appearing knowledgeable, confident, and put-together. Scientists like to believe that they are judged by their results and not their jeans. But remember, this book is not about how to talk to your peers and academic professors. It is about dealing with decision makers, nonscientists, and experts in other fields.

Your clothing does matter. It should be professional, fit properly, and be appropriate for the occasion. In a formal presentation setting, it's always best to be a little more dressed up than your listeners. Yes, that may mean sporting a tie! If they will be in business casual attire, wear a sport coat, suit jacket, or dress and jacket. If they will be in suits and ties, wear your holiday best. Make sure the quality and condition of your shoes wouldn't embarrass your mom. It's human nature to notice everything a speaker is wearing—especially when they are on stage talking for twenty minutes or longer.

Don't wear anything distracting. You want to keep the audience's attention on what you are saying, not the fact that your shirt is too sheer, your pants are wrinkled, or your jewelry is big and bright. If you have lost or gained weight recently, double-check the length of your pants, as they can become too short or, worse, puddle around the top of your shoes. Choose colors that flatter you (ask your friends) and patterns that are subtle and understated. A bit of color in a tie or scarf is fine. Open-toed shoes are a bad idea, and so are short skirts, shorts, or clothes that are tight fitting or call attention to your legs. When you look in the mirror, you want to see a subtle canvas that flatters you without calling attention to your body or your fashion sense. If you tend to get visibly nervous when you speak, wear dark clothes that won't let anyone see you sweat—literally.

As you read these paragraphs, you may be thinking, "Hey, what about that really famous guy in my field who looks like the cat dragged him in? He does fine, right?" Well, he undoubtedly built his reputation on solid research and publications and not his disheveled appearance. People like this must invest a certain fraction of their influence in asking the listener to believe them even though they don't match the traditional view of what a credible scientist looks like. So be very careful of playing this game—it is best left to those who have reputation to spare.

Formality standards in dress have shifted in the last few decades. There are environments where CEOs come to work in jeans and decision makers dress casually. That might tempt you to mirror their attire. You can certainly opt to do so, but we advise that you wait until you have built your credibility and established a relationship. Err on the side of dressing up when making a first impression.

PRACTICE YOUR FIRST TWO MINUTES

Getting your opening right takes practice. Recruit a trusted colleague or two to listen to you and provide feedback. Include someone who is not intimately familiar with your subject matter. Help them understand your audience and your goals. Ask them if your approach has captured their attention. Have you made a connection? If not, ask what you could do or say to accomplish that objective. Probe for feedback by asking about the use of specific words or phrases. Reiterate that you want constructive ideas. See if they can tell you what the essence of your message is. If listeners can summarize and repeat it back to you, you know you are being heard. Often people are reluctant to tell you what they really think unless you make it clear that you truly want their candor. Tell them you need their help so you can become more self-aware and self-correcting. Accept their input willingly and ask questions to clarify what they mean if you are unsure. Their input will help you hone your message.

7

Crafting Key Messages and Narratives

How many times have you endured a painstakingly detailed discussion or presentation where a scientist trudged through their entire journey of discovery to show you how they derived the answer? Too many? That's not a surprise, really. Bill Young recalled from his days as an engineering student at Purdue that there was no course or seminar on how to give a compelling talk. When we asked Bill about how he was taught to communicate in college, he said, "I remember that very well . . . because we weren't. I was an English minor, but you have to remember that engineering school is pretty intense. The courses are crammed with math and physics and chemistry, and you don't have a lot of extra time or room for communication skills."[1]

After graduating, he found the realities of communication to be quite different: "You get out into industry, and you find out, 'Oh, you actually have to communicate your ideas.' I quickly learned you have to break complex science down, especially when you talk to nonscientists. You have to be able to work in teams. None of those skills were taught, and I don't think they are to a large degree today."

Bill chose the pharmaceutical industry and took a position at Eli Lilly in Indianapolis after graduation. His big break came following an assignment in Puerto Rico, where his team transformed a bankrupt brewery into a plant for manufacturing antibiotics. Bill was recruited back to Indianapolis to serve as the technical lead for a new collaboration credited with the first recombinant, genetically engineered product—human insulin.

Lilly's partner was a fledgling company by the name of Genentech. In 1980, Bill joined Genentech about a month before its $35 million IPO—which is well remembered for its meteoric rise from $35 to $88 a share after less than an hour on the market. Bill spent twenty years at Genentech. He was initially charged with developing the processes for making new products and ultimately went on to become the company's chief operating officer in 1997. Seeing the early promise of personalized medicine, he left in 1999 to become CEO of Monogram Biosciences, a diagnostic company addressing viral disease and cancer. About ten years later, Bill sold the company to LabCorp and joined Clarus Ventures to focus on investing in promising healthcare companies.

Bill often witnesses the lack of communications training when scientists approach Clarus Ventures for funding. He told us that most seem to follow a basic tenet that goes something like this: "Since I went through years of hard work and hell to figure this out, I'm going to take you through the whole thing, and at the end I'll show you how I got the answer." He advises scientists to "give the answer up front and then, if we're interested in how you got there, we'll ask." He recognizes that "turning it around" is difficult for many scientists but says that it's often the best way to start.

Nancy Floyd, founder of the venture firm Nth Power, concurs: "I'd say that 90% of the time, we get inundated with a lot of technical detail. In some ways that's understandable coming from technologists and scientists. But that's a big mistake. There will be plenty of time during due diligence to get into the technical details. But first, you've got to hook the investors."[2]

"Look at how much work I did" is never the story or the message you want to convey. Yet all too often, that is what comes across, in large part because scientists don't pay attention to creating compelling narratives. Fellow scientists may relate to the detailed challenges you've faced, but when you are presenting to decision makers, it's more important that you get them to care about your science. You want to leave them feeling *confident in you, excited about your topic, and able to take action.*

So here are three essential elements to think about as you craft your content: the setup, the key messages, and the story. These will help bring your science to life in a memorable way.

ELEMENTS OF A COMPELLING DISCUSSION

The Setup

Just as in the five slide approach (chapter 3, "Extracting the Essence"), start with the problem we face today that needs a solution.

What's the current situation? Tee up the problem and give a brief explanation of how it is being addressed.

State why today's solution falls short and what has to be done to fill the gap.

How will you solve it? Focus on why your science matters and why listeners should care.

Keeping this introduction short is hard, especially for scientists who value precision and accuracy. But this is simply the setup. Instead of giving comprehensive descriptions and overly detailed hooks, you can say things like, "What if we could turn plants into plastic?" or, "What if we could run everything without gasoline?" Your fundamental point should be that there is a better way or a different way. You don't need to belabor the argument by listing the merits and shortcomings of each of the current approaches. Your goal is to get your listener primed for the message about your science.

As you are fleshing out how today's solutions fall short, write out what comes to mind. If it is complex, force yourself to take it up a level. Forego some of the detail. Consider using iconic analogies that will help listeners more quickly grasp a complex foreign concept. A familiar mental image will immediately ground their understanding of what you are doing—it places your science in the context of their life experience instead of just within the sentence of words you are speaking. And remember that at this point in the discussion, it is not extreme precision you are looking for. You just want them to follow your point.

Here are some examples of iconic analogies in use. To describe what a national lab does with prospective partners, Julio Friedmann uses the image of the space shuttle lifting off:

I point to the booster rockets and say, "That's the national lab," and I point to the shuttle and say, "That's you." You have a mission and the national lab's job is to help you get your mission done. It is an iconic image and it helps people understand the essence of the relationship.[3]

In his Three Minute Thesis (see "Examples of Great Openings" in chapter 6, "The First Two Minutes"), Dustin Chernick, a pharmacology student at the University of Minnesota, discusses how the brain uses good cholesterol to reduce the buildup of toxic plaques that contribute to Alzheimer's disease. He explains that glial cells can normally clear away these toxic proteins faster than they get produced, kind of like a sponge. He extends the sponge analogy with a kitchen reference:

> If we think of the brain like a kitchen, when we make dinner, we make a mess, but we clean it up with a sponge. As we age, and our good cholesterol starts to function less effectively, these cells lose their ability to clean up the toxic proteins fast enough. The sponge gets grimy. But if we could clean the sponge, then we could get the kitchen back in order. My research is to find out if we can use a drug that mimics good cholesterol to clean the sponge and boost the brain's ability to clear away toxic proteins, thereby rescuing memory function in Alzheimer's disease.[4]

Suryashree Aniyan, a very creative student from the Australian National University college of Physical and Mathematical Sciences studying dark matter, uses the iconic analogy of marshmallows floating in hot chocolate to open her Three Minute Thesis talk and engage her listeners. She likens watching marshmallows bob up and down to measuring the wobble of stars to estimate the amount of dark matter in the galaxy: "In thick, delicious hot chocolate, the marshmallow would move slower than it would move in diluted, watered-down hot chocolate."[5]

One note of caution: don't get carried away or become too cutesy. Suryashree's marshmallows and hot chocolate analogy might be too frivolous for the opening of a talk at a scientific conference or in front of a science sponsor, but it works well as an opening for her college competition.

The Message

Key messages are the carefully selected phrases that extract the essence of your science. Choose language that is clear, memorable, and if possible, attention-grabbing. At a minimum, make sure it's easy to follow. These are the statements you use over and over again in every conversation and presentation. Whether you are writing, chatting in a hallway, or having a beer, stick to similar phrasing. You should become tired of saying these words. Use them exclusively and consistently to make sure listeners hear what you view as the important concepts, conclusions, and considerations.

Why Are Key Messages Important?

Key messages are like the refrain in a good song. They set up and organize the rest of your content. They are the embodiment of extracting the essence.

Decide exactly what you want your listener to know and focus on. Include only the information that sets up why your science has value, and remember that your goal is to compel listeners to take action or agree to your ask.

The discipline of developing key messages also creates a practice of prioritization. It forces you to decide the three most important things you want people to know and remember. When you have three key messages (considered to be the maximum number most people will remember*) you naturally end up arranging them in a hierarchy. Simply put, key messages help you stay on track and communicate your science with more impact.

*For a simple introduction to this issue, see Susan Weinschenk's book *100 Things Every Presenter Needs to Know about People* (Berkeley, CA: Pearson New Riders, 2012).

Limit yourself to creating no more than five, but ideally three, messages to be certain your key points stand out. The art for a scientist is learning to reinforce key messages by returning to the same ideas without flat-out repeating them (something that is expected in business but can insult a technical audience). Incorporating proof points in the form of data, stories, and examples that support and illustrate your messages enables you to reinforce without repetition. This is a key difference between the conventional science talk you give at a university seminar and the kind of communication we are describing.

Developing and delivering key messages makes your communication purposeful. To borrow a phrase from Stephen Covey, author of *The Seven Habits of Highly Effective People,* when you craft a conversation or presentation you want to "begin with the end in mind."[6] This approach will ensure you have a focused discussion designed to enable your listeners to understand what is most important and the reason you are communicating with them.

The Narrative

Provide the highlights of your past struggles and current challenges. Keep it short and sweet. What options have you weighed? Did anything surprise you? How did you choose which direction to take? What happened next? Which lessons did you learn? At this point, you are the hero taking us along on the adventure of advancing your science. Draw your listener into the narrative by discussing situations that they too may have experienced so they can relate to you.

THE SHORT SETUP: A CELLULAR TECHNOLOGY EXAMPLE

Despite the fact that early predictions for the number of wireless users were off by literally millions worldwide, industry leaders and technologists knew that analog network overcrowding was inevitable due to radio spectrum and channel constraints. In the early 1990s, during her days as corporate spokesperson for AirTouch Communications, Amy was embroiled in explaining the emergence of new digital wireless communication technology. The analog capacity problem spurred the development of two competing, incompatible digital cellular technologies, TDMA and CDMA. TDMA stands for *time-division multiple access,* and CDMA is short for *code-division multiple access.* Both methods significantly increased the quantity of concurrent users within a finite section of radio spectrum. At the time, this was a much-needed technological breakthrough.

In simple terms, the two approaches had different specifications for how the bandwidth of the radio spectrum was distributed during cellular calls. TDMA divided the channel into sequential time portions. Users of the channel would take turns receiving and transmitting call data. Only one caller could actually utilize the channel at any given time, sending information in short bursts. Meanwhile, the other callers momentarily gave up their use of resources to enable that caller to use the channel. Imagine a relay race where one runner doesn't start until the one in front of them is finished. (The relay race example is an iconic analogy.)

In contrast, CDMA enabled numerous callers to use the channel simultaneously. This was made possible by *spread spectrum modulation,* which captured digital bits of every conversation and spread them all around the channel in a pseudo-random manner. At the receiving end, the scattered bits were reassembled in proper order to make them coherent. CDMA allowed multiple conversations to be carried out via the same channel simultaneously. TDMA did not.

But in the early 1990s, TDMA was operational, while CDMA was not ready for prime time. Proponents of CDMA, including Amy's company AirTouch Cellular, were making the case that this superior technology was worth the wait.

Why all the explanation of digital wireless technologies? To give you some background for the following example of a short setup. Here's how Amy would set the context for why CDMA was worth the wait when talking at a wireless event to a group of cellular industry veterans who already knew that cellular networks were overcrowded and network quality was starting to suffer:

> As of October 1993, about twelve thousand people were signing up for cellular service every day. With overcrowded networks and no end in sight to such rapid growth, wireless carriers need a digital technology that can make the most efficient use of the radio spectrum. CDMA is worth waiting for. It will have more capacity than analog and more capacity than the current digital technology, TDMA.

Notice that Amy did not dive into the details on how the two technologies work. She is still in the setup stage and doesn't need to provide that detail. She simply needs to set up the discussion to focus on the merits of CDMA.

She could have been more detailed and said, "CDMA is worth waiting for because it uses spread spectrum modulation, which gives it more capacity than analog and TDMA, the first digital technology." Doing that would layer on something else for her listeners to grasp. It would make her statement more complex. It would distract from the focus she wants them to have, which is the case she is about to make for what this superior technology delivers. Her message is succinct: CDMA is worth waiting for.

Exactly what do you want your listener to know and focus on? Be relentless about *leaving out* details and background that detract from that focus. Keep asking yourself questions: Why does my listener need to know this? Will it get them to care? Do they need to know it now or can I provide the details later? Will it compel them to take action or agree to my ask? Resist the temptation to share all of what you know to show your command of the subject. A cogent discussion is the best way to engage your listeners. Keeping it simple and consistent is the goal.

THE IMPORTANCE OF CONSISTENT MESSAGES

Jay Davis (of the five slide approach; see chapter 3, "Extracting the Essence") reminds us how much sending a consistent message matters when you are addressing multiple audiences and building support for your ideas. He saw this play out while working at the Defense Threat Reduction Agency (DTRA):

At DTRA, you learn that you've got to be on message and absolutely say the same thing all the time. If you try to send a different message to a different constituency, those lines will close behind you so fast your credibility goes to zero. I was better off to look at somebody and say, "You're not going to like what I'm going to say but this is my position on this," than to try to tell them something that makes them feel good, and have them discover that I said something very different two offices down. And you have to be infinitely sensitive to the fact that anything you write or anything you say can instantly be on the front page of the *Washington Post* the next day.[7]

Here is a messaging example with proof points from the project Jeff Johnson (the guy who has mastered the art of warming up the room; see chapter 4, "Who's Listening?") was championing—FirstNet, the first nationwide wireless data digital network dedicated to public safety. An important setup for the key messages that he used with every audience was, "Believe it or not, our public safety community typically uses their personal smart phones instead of officially issued equipment because they have more applications and better coverage and functionality using their own devices on the same digital wireless network as consumers."[8] This helped explain the problem and why there was a need for a brand new nationwide digital network dedicated to public safety that, for the first time, would provide state-of-the art service to police, fire, and public safety personnel so that they wouldn't have to share commercial wireless networks.

The communications team that Amy was on developed two over-arching key messages (shown in bold below) to introduce FirstNet and position it as a must-have technology for public safety professionals. They also crafted the support point bullets that specified how the advanced functionality of this dedicated network would be better and distinct from the existing public safety networks. They used these messages every time they communicated about the promise of FirstNet. The messaging strategy summary was:

MESSAGE I	MESSAGE 2
Dedicated to public safety	**Reliable for public safety**
Proof Points	Proof Points
• Dynamic priority access	• Seamless connectivity
• Incident command control	• Tailored coverage
• Local management	• Cyber security
• Lower costs	• Public-safety grade

TABLE 1   MESSAGE PLANNING MATRIX

Segment	Think (current point of view)	Feel (level of interest, emotional connection)	Do (action you want your listeners to take)	Shift needed from listeners
Funder/Sponsor				
Regulator				
Supporter				
Collaborator				
Competitor				
Naysayer				

PLANNING YOUR KEY MESSAGES

There were numerous opportunities to talk about FirstNet in a multitude of forums addressing audiences ranging from fire, police, and emergency medical technicians to governors, Washington policymakers, wireless industry companies, and state telecommunications management. When you have diversity like this, how do you tailor your messages? One safe bet is to craft your key messages based on what would appeal to an end customer or recipient of the benefits of your science. That way, you are likely to find a message that will resonate with everyone else—even naysayers. Try to get inside the hearts and minds of your listeners. What are they likely to think about your topic? What are they likely to feel? What action do you want them to take as a result of listening to you?

As you strategize about your messages, you might find it helpful to create a matrix that looks at major audience segments with an eye toward assessing their current frame of mind (see table 1). Putting yourself in their shoes will help you think about how best to approach, influence, and persuade them.

COMPLETING THE MESSAGE PLANNING MATRIX:
A BRAIN TRAUMA EXAMPLE

Eric Blackman is a professor of physics and astronomy at the University of Rochester. Among his interests are the physics of brain injuries. In his presentation "Helmet Protection against Traumatic Brain Injury: A Physics Perspective,"[9] Blackman includes ample scientific detail. His presentation describes chronic traumatic encephalopathy (CTE), a term you might recognize if you happened to see the 2015 movie *Concussion*.

The film depicts the fight of the brilliant forensic neuropathologist and Nigerian immigrant Dr. Bennet Omalu (played by Will Smith) to get CTE recognized in the scientific community. He deemed it a football-related brain trauma. He found that the average football player sustains an estimated 950 impacts to the head during a season. These hits could result not only in concussions but also in long-term brain damage.

In his scientific presentation, Blackman sets up why we should care by explaining what happens in the brain as a result of CTE. In patients suffering from CTE, a toxic tau protein builds up in brain cells, preventing normal connections to other cells and causing cells to die. Blackman's presentation dives into the jargon of the problem for a moment, discussing how tau protein shows up as neurofibrillary tangles and glial tangles. Tangles are formed by hyperphosphorylation of tau proteins in microtubules, causing tau to aggregate. These phenomena accompany dementia, although they are not themselves a signature of Alzheimer's (primarily because there is no beta amyloid present). CTE is prevalent in the brain tissue of deceased football players and boxers, even some who don't have a clinical history of excessive concussions.

Below are some of the key conclusions that one could draw from Blackman's presentation. These are the kind of statements that work well as key messages. These statements can be supported with the data from Blackman's research. In the slide presentation he created, Blackman has the opportunity to reinforce the messages with his headlines and the way he narrates his slides.

Three key messages are:

- Hard shell helmets alone won't prevent devastating head injuries.
- Data shows that proper cushioning reduces head impact acceleration and thus force on the brain.
- Cushioning standards must be more stringent to protect against closed traumatic brain injury and do more than just prevent skull fracture.

Taking Blackman's science as the topic, let's look at a message planning matrix that examines the National Football League's 2016 plans to test out the new ZERO1 helmet, a $1,500 flexible helmet codeveloped by the University of Washington and the Seattle-based start-up VICIS. Let's say you work for VICIS and are speaking at a conference of football team physicians and neurologists. Table 2 shows how you might populate the message planning matrix to prepare for your talk.

TABLE 2   COMPLETED MESSAGE PLANNING MATRIX

Segment	Think (current point of view)	Feel (level of interest, emotional connection)	Do (action you want your listeners to take)	Shift needed from listeners
Funder/Sponsor	Helmet technology is promising and solves an important problem.	I want to invest because I see value in helping athletes protect themselves.	Consider investing.	Get them to understand the value of the product.
Regulator	Better helmet technology is essential. Current helmets fall short.	I want to see the results of the pilot.	Require the results of the pilot to be made public and compare them to current helmets.	Get buy-in to report pilot results broadly.
Supporter	Better helmets are necessary.	I want to see this pilot succeed.	Recommend the use of the new helmet.	Enable them to make the case for better helmets.
Collaborator	Together, we have developed a breakthrough solution.	I am excited about the pilot.	Actively support the pilot efforts.	Identify a clear way to play a collaborative role in the testing.
Competitor	I have a better solution.	I don't want the pilot to succeed	Be respectful of your new helmet.	Refrain from bashing your product.
Naysayer	I don't think this new helmet technology is necessary.	I don't see a need for this technology.	Be open to the benefits of the new helmet, be convinced it matters.	Acknowledge better protection is worthwhile and that athletes deserve protection.

Once you have done this homework, you are ready to craft your key messages. You now understand the potential attitudes of your audience segments. From here, you can figure out the kind of call to action that you can craft to be inclusive. You want to ensure that the points you make will address the action you want people to take. Arm your listeners with information they need. And don't fall into the information deficit trap that we discussed in chapter 5, "Why Scientists Communicate Poorly outside Their Field." Key messages are not meant to convey large volumes of information. They are intended to specifically address the listener's concerns, biases, and your communication goals.

The hypothetical football helmet presentation is rich with opportunities to cite statistics about head injuries and the number of ensuing deaths. You could find an analogy to help your audience imagine the force and impact of a hit to the head. You would want to explain the helmet technology with an emphasis on how it protects the brain from sustaining injury. You could even show a clip or paraphrase scenes from the movie to illustrate how the NFL didn't want to believe there was a problem. These elements, woven into a conversation, would certainly get everyone listening.

The concluding call to action could sound something like this: "It's time to do right by athletes young and old and protect them from head injuries. We know how. Hard shell helmets alone don't cut it. We need your support to trial new technologies with better cushioning, like our ZERO1 helmet. We need more stringent standards to protect against chronic traumatic encephalopathy."

CREATING ENGAGING NARRATIVES TO CONVEY YOUR MESSAGE

Sharing the highlights of how you made your scientific discoveries is a compelling way to drive home your message. Stories are engaging. Stories are memorable. People make sense of the world through the structure of stories. The kind of stories that work for scientists are narrative, factual explanations. Kendall Haven, an oceanographer, West Point graduate, and former Department of Energy scientist, now dedicates his career to helping scientists communicate. In his book *Story Proof: The Science Behind the Startling Power of Story,* he says, "The use of story structure as a way to organize and present material is more effective simply because it matches the way that humans naturally think and perceive."[10]

Haven explains that it is vital to reveal the struggle: "Science doesn't happen by itself. The people who do the work, their challenges and struggles are the story. Tell listeners about the people in order to make the science comprehensible and meaningful." He adds, "When the topic is unfamiliar, familiar story structure becomes a valuable form of prior knowledge to activate in order to guide the listener toward creating meaning and memory." Well-told stories bring the listener in to share the experience. Stories get retold. Stories capture hearts and minds. They can provoke an emotional response and compel people to take action.

As uncomfortable as this is for a scientist trained to talk in the third person, telling *your* story is a fabulous way to engage listeners outside your field.

- Start the narrative by establishing the setting and circumstances.
- Introduce the main character (you or your team).
- Describe the challenge.
- Explain how you addressed the challenge and what happened.

Your narrative will be more memorable if you create pictures with your words. You can do that by giving some colorful details, especially about changes in direction: "We pursued this line of evidence for two years, but it just didn't pan out. Then my postdoc Tom noticed this new effect—and we had a success." Consider what you want the listener to learn or do as a result of hearing the story. Be sure to drive that point home throughout the story and again at the end.

A COMPELLING EXAMPLE FROM HEART HEALTH

Cardiologist Dick Friedman knows how to tell a compelling story when it comes to championing the use of statins.[11] Dick does not cite numbers. Instead, he shares his personal experience of treating heart attacks (myocardial infarctions to Dick's medical colleagues) in the emergency room.

When his pager wakes him up in the middle of the night, it is often to respond to a heart attack. He has to perform a cardiac catheterization procedure within ninety minutes to minimize damage to the heart. The attack starts with a small tear or disruption in cholesterol deposits lining the artery. In response to the damage, clotting begins at the site of the tear. The combination of clotting and the rupture of the cholesterol plaques can quickly block the artery. When blood flow to the blocked portion of the heart stops, damage to the heart muscle begins. The faster the blood vessel can be opened by the cardiologist the better. Time is muscle damage. The ideal result is no damage at all.

Most of those crucial ninety minutes are used in discovery—getting to the hospital, getting the patient's heart restarted with heroic emergency room procedures if necessary, getting the test results back that say the stoppage was caused by a blockage and not a drug overdose or electric shock or any of the dozens of causes that the emergency room must consider. Myocardial infarctions, and ultimately the short window of time to begin the catheterization procedure, are the reasons Dick is so attentive to his pager.

But the urgent response of the trained medical team is not the focus of Dick's story. Having described the events in the artery, in the hospital, and in the minds of the medical staff diagnosing the cause of the event, he abruptly stops his story. He proclaims that only once in his experience has a heart attack started that ninety-minute clock of life in a patient taking statins to reduce cholesterol levels.

The message is clear. Dick thinks you should take statins to treat high cholesterol. His narrative is compelling. But not because of the statistics, which are so clearly in favor of the medication. It is powerful because it is the personal story of a man who is dedicated to putting himself out of business by conveying clear medical advice in ways that people can understand. He has iconified our fascination with *M*A*S*H* and *Grey's Anatomy* and conveyed the benefit of taking action—in this case, taking the medication—in terms of his own experience. He tells patients that taking statins is like wearing your seatbelt and your shoulder harness while driving a car equipped with airbags. Statins dramatically lower your risk of having a heart attack. He has iconified the statistics with his own decades of medical practice. And Dick uses those icons to tell the story in about five minutes, compelling the patient to take action.

Decision makers who are in a position to provide funding often invest in *you as a person,* not just in the merits of your science. Dick Friedman is believable because he puts himself in the story. Decision makers want to know who you are and whether you are worth the investment. But they didn't get into positions of authority by being susceptible to hyperbole and kidnapped credit. Be completely honest about what you have done and how you plan to do even more. This is science, and it is your responsibility to tell both a good story and an accurate and honest story.

TELLING YOUR STORY

We hear a lot about the power of using stories. There is enough of this in the popular literature to make any scientist question the validity of the claims—and doubt that it really applies to science. But humans really are "wired for story," which is the title of Lisa Cron's excellent book on the subject, *Wired for Story: The Writer's Guide to Using Brain Science to Hook Readers from the Very First Sentence.*[12] From the earliest days, humans have transmitted information in the form of stories, and good stories have a distinct structure that utilizes all those generations of hardwired brain structure. The end result is that your entire brain participates when you hear a good story and you simply absorb information faster.

It is easy to understand why podcasts and TV movies take advantage of your brain's affinity for story. But this is not part of the typical academic talk structure, is it? You will find that it can be used to great effect. Much of the story, however, is prewired in your audience's *experience*. Your academic colleagues know why the problem you are attacking is important—they may even know why you decided to pursue it. They know who has tried and failed in the past. There is more of a story there than you may realize, and while your scientist colleagues may know it, most decision makers will not. So tell it.

Use Story Structure

Speakers who really hold your attention take advantage of story structure, often unconsciously, to make science more interesting. And speakers who are talking with nonscientists, especially the decision makers this book is focused on, know that they have to take advantage of those prewired pathways to keep their listeners' interest high and to help them understand what can be a complicated tale. As Lisa Cron states, "From the very first words, the listener must want to know what happens next."

Lead with the Setting

Describe the time and place where your story begins, as Dick Friedman did by beginning his story asleep in bed. Use sensory details freely to draw people in so they feel they are right there with you, sharing your experience. You might open with a line like this: "I was sitting in the lab on December 21st watching the sky grow dark after eleven hours of running the experiment and feeling the pressure of the holiday, just days away, when . . ." Engage your listeners and evoke their empathy. A *lot* of this in a science talk is maudlin, but a *little* of it is gold.

Introduce the Hero

Here's your opportunity to bring to life the people that are on your team and to establish your credibility and expertise through the way you tell your story. Give us a sense of the personalities and nature of the main characters. Be sure to convey what motivated the team to do their work and what they wanted to achieve.

Reveal the Struggle

Highlight the challenges and problems that you and your team have overcome. Tell the CliffsNotes version—don't discuss every setback in detail. Choose one to amplify to help bring the audience along with you.

Bill Young told us the story of the challenges of bringing new scientific ideas to market and the good luck that ushered in Genentech's first oncology drug—Herceptin:

> The whole experience at Lilly and Genentech was almost out of a fairytale. We had great times and there were ups and downs. It wasn't always an easy ride. In the early days at Genentech, we thought that biotech could do everything, so we had an agricultural division and we had an industrial division. After a while we decided that making drugs was pretty time consuming on its own, so we formed Genencor and another company to do the industrial work and licensed off the agriculture piece. We ultimately focused on drugs, and in time, on cancer, which happened rather serendipitously. Herceptin was the first oncology product, followed by many other anti-cancer products targeting specific receptor-site cancer cells.[13]

He continued, introducing the characters and heroes in the story and revealing their struggle:

> In the mid-'80s, one of our best molecular biologists, Axel Ullrich, working with Art Levinson [who later became the company's CEO], cloned the first full-length human HER-2 gene. We didn't really know what to do with it back then. We cloned this receptor, filed a patent and put it on the shelf. Then Denny Slamon from UCLA observed that women with breast cancer who overexpressed this receptor were at risk for rapid progression of their disease. About 25 percent of patients have this overexpression of HER-2. He came to us with this data and said, "Look, you really should do something

about this." He talked to everybody in the company trying to get people to pay attention.

The atmosphere then wasn't conducive. We had just had some disappointment with our TPA (tissue plasminogen activator), and most of the focus had been on that cardiovascular product. Cancer was not as fashionable as it is now. In fact, we had done a cancer trial that had failed. Targeted therapy was a brand new idea. No one had done it before. On top of that, you had to use monoclonal antibodies to attack the target and that was fraught with challenges. You needed to humanize them and make them cheaply enough in large scale. It took a lot of protein to have a decent product and a decent cost margin. There was a lot of discussion inside the company, is this the right product? We couldn't afford to do them all. There were not a lot of oncologists at Genentech at the time either.

Finally, Bill personalizes the story by sharing his mother's experience, absolutely making us care:

We had many discussions with Slamon. I recall one where he was sitting in my office with Mike Sheppard, one of our molecular biologists. Denny showed me the data on what happened to women who overexpressed this receptor. Mike showed me data from the laboratory—mouse versions of Herceptin, early versions of the product with lines of cancer cells. For me that was a seminal meeting.

Part of the story was that my mother was diagnosed with breast cancer at about the same time. I remember talking to oncologists about how few treatment options she actually had. Most of the ones she did have made her sick and really didn't do much for the disease. We all saw there was a huge need there. We believed we were ushering in the new science and we ought to be able to do better than the current technology. From then on I was an early supporter.

(If you want more details on this fascinating story, take a look at Robert Bazell's book *Her-2*.[14])

This is the fundamental difference between the talk you give to your professional society and the conversation you have with decision makers: your colleagues know why they care about your research because they have heard much of the backstory before, often during years of study. But decision makers need to know why your story matters to them.

One very practical book, which also happens to be Roger's favorite book on storytelling, is *Connection: Hollywood Storytelling Meets Critical Thinking,* by Randy Olson (a PhD marine biologist) and his collaborators Dorie Barton (an actress) and Brian Palermo (an improv instructor).[15] Olson reminds us that simply reciting information is not the way to start a story. Your data is important, and while he acknowledges the need for a factual basis for discussion, he says that's not enough. According to Olson, "A story begins when something happens." Just listing facts and findings *does not* constitute something happening.

Olson points out that movies today have much less lead-in to the action. Perhaps this is because we have more shared experience to draw on—when the hero gets into the car for the chase, we all understand what cars are and the basic physics behind the chase.

Listeners get restless when the lead-in gets too long. People expect the story to start. The character issues and background can be exposed later, after the audience is sure it's a story they want to listen to. People have to care to want to follow a story—whether it is a movie or a grant request. "Why do I care?" This is the question you must keep your listeners from asking—let them know why they care.

THE AND-BUT-THEREFORE MODEL

Olson gives a simple formula for creating an interesting sentence from the typical list of "this, *and* this, *and* this, *and* this" that scientists often have in their heads. Replace *and* with *but* or *therefore.* In its most basic form (which he attributes to Trey Parker, the cocreator of *South Park*), Olson's version of the scientist's story is, "This *and* this are true, *but* this flies in their face, *therefore* the following things must be done." "This *and* this" represent the exposition, the preliminaries, the setup for telling the story. Olson argues that scientists often never get out of this mode—they use their entire time on exposition and don't get to the point.

Olson may be a little harsh on scientists here, but the criticism is fair. Scientists often spend too much time showing result after result without providing a good narrative. How many talks can you remember where there was no exposition of a fundamental gap in our knowledge or a fundamental need for new technology? Great talks have them—and persuasive speakers are not afraid to trot these points out early and then spend most of their time explaining what they, or you, can do about it. This is the *therefore* part of the talk. The same general principles apply in conversations. Just be sure to keep your story short.

NARRATIVES WITH IMPACT

General stories are vague, and vague stories are boring. Olson cites the old adage, "One death is a tragedy; one million deaths is a statistic." To capture the hearts and minds of many decision makers, the best narratives are those told in detail about a specific situation. Dick Friedman's story is compelling because it is about one doctor's struggle with heart attacks.

For example, you could tell the story of a girl with a fatal disease, one million people with that fatal disease, or a girl with a fatal disease shared by one million people. Olson contends that the best *story* is the first one and that no public telling will get more attention than the single-victim version. He calls that "the power of one." This is almost certainly true for moviemaking and capturing the attention of the general public, but it points out the evolution that you must make from scientist to champion. You probably work on that disease *because* there are a million cases of it, but if you want to advance science in the ways we are talking about in this book, you *also* need to paint a picture of the large-scale problem in more focused terms that decision makers can absorb.

You might use the story of a single patient as an opening to discuss your newest finding, as we saw in the winning Three Minute Thesis talks, or midway through a presentation as a way of reinforcing why it is critical that your research be funded. We are not saying that you should leave your data behind and only tell the "power of one" story. We suggest doing both.

The Yarnell fire tragedy told in chapter 4, "Who's Listening?" in which nineteen of Arizona's Granite Mountain Hotshots lost their lives, is a good example of this story power. Although many more (343) firefighters died in the 9/11 attacks, the Yarnell story stands out because it vividly referenced the impact the disaster had on a single close-knit team of heroic firefighters. The death of those men clearly could have been avoided by new technology. You feel their loss in your gut when you hear the story.

8

The Power of Language

Finding the ideal words to talk about your science takes time. Your goal is to arrive at phrasing that is clear, compelling, and memorable. Then keep refining it to be as succinct as possible. You don't have to get carried away with hyperbole. Simple, straightforward language works best.

On their homepage, ORIC Pharmaceuticals describes their mission in two very different statements. We find the company's official mission statement, "Passionate about discovering meaningful therapies for patients," far less compelling than the banner that appears on the same page: "Our mission is to overcome resistance in cancer."[1] The phrase "overcoming resistance" immediately tells us they are solving a difficult problem. The statement is clear and simple. And by the way, ORIC is an acronym for Overcoming Resistance in Cancer.

In your search for the right words, we recommend that you test your story on listeners who don't know much about your topic. Ask them to tell you what they take away from what you've explained to see if they are following you. Figure out how to clarify points of confusion. This feedback loop will help you be self-correcting so you can be better understood.

ANALOGIES PROVIDE AN ANCHOR TO THE UNFAMILIAR

One very effective way to create a bridge for understanding new or complex concepts is to use metaphor and analogy. When your science is unfamiliar to a decision maker, they are not likely to have sufficient background to make sense of technical details. The right metaphor or analogy can enable them to apply their experience with something familiar, which becomes a steppingstone for grasping the new concept you've introduced.

In episode 3 of *Emperor of All Maladies,* a PBS documentary about the history of cancer, the narrator describes tumor suppressor genes as the opposite of oncogenes. He then says, "Oncogenes work like an accelerator in a car—they drive cell growth. Tumor suppressors are the brakes. In many cases, brakes are simply gone from the cell." Robert Weinberg, a professor of biology at MIT, follows, explaining: "We now realize that virtually all human cancer cells have both defects . . . *stuck accelerator pedals* (hyperactive oncogenes) and *defective brake linings* and together those two defects conspire to make the full panoply of behaviors that we associate with malignant cells."[2]

The beauty of using familiar concepts like a stuck accelerator pedal and a defective brake lining is that the listener doesn't need to know the biology behind the terms *oncogene* or *suppressor gene.* We immediately understand that a stuck accelerator means the disease is out of control and gaining momentum. Likewise, we recognize what's at stake when brakes fail. Using this language simplifies the listener's task and helps them understand the mechanism being discussed.

The FrameWorks Institute is a nonprofit that "designs, conducts, and publishes multi-method, multi-disciplinary communications research to empirically identify the most effective ways of reframing social and scientific topics . . . and offers strategic guidance and a variety of professional learning opportunities for advocates, scientists, policymakers, and nonprofit leaders." In its report *How to Talk about Climate Change and the Ocean,* FrameWorks uses the explanatory analogy "heat-trapping blanket" to help people understand how climate change works because it has consistently proven to be a reliable tool for expanding public understanding:

> Heat-Trapping Blanket: When we burn fossil fuels for energy, such as coal, oil, or natural gas, we release carbon dioxide into the atmosphere. Carbon dioxide is a gas that traps heat. As CO_2 builds up, it acts like a blanket, trapping in heat that would otherwise escape. This "blanket effect" is warming the planet's atmosphere, disrupting the balance that keeps the climate stable.[3]

What is your equivalent of a heat-trapping blanket or a stuck accelerator pedal? Can you zero in on an aspect of your science that has a parallel to something familiar? Even if you start with one analogy that covers one aspect of your work and expand from there, you have made a meaningful step toward helping your listeners gain easier access to your science.

AVOID USING THE WRONG WORDS

Scientists should avoid some words and phrases because they can cause the listener to disparage rather than value the message:

apparently; obviously; decidedly; basically; clearly	Avoid these. If any of these words are true, let the audience deduce it. This lets the listeners become part of the scientific process and avoids making you sound like a county fair huckster.
to be honest with you; candidly; the fact is	These words imply that you have failed to be honest with your listeners up to this point, putting them on high alert for dishonesty and eroding trust. Your goal is to build credibility as an expert, so it's best to steer clear of these phrases.
personally	This implies that you have a different opinion than your employer or sponsoring institution—acceptable at lunch, but not in a presentation.
It is what it is.	What does this mean? It implies that you have no idea why something is happening. It has no place in scientific discourse.

Other words and phrases come into our speech patterns from our social interactions, but they can convey the wrong message when used in a presentation. Based on the system of Strunk and White,[4] the examples on the left are to be avoided, while those on the right convey the idea more appropriately.

Wrong: Play with the data.	Better: Examine how the results are affected by changes in the data.

The form on the left suggests that random changes are not only acceptable but expected. The form on the right suggests that there is a range of possible behaviors but that you are systematically evaluating them.

Wrong: I just made these slides last night.	Better: These are early results.

Rather than making your audience feel like you didn't care enough to thoroughly prepare for them, let them know that they are the first people you wanted to bring into the discussion.

Wrong: This is the talk I gave for X.	Better: I shared these ideas with X and thought you might like to see them.

This should be used only when circumstances unavoidably make it obvious that your presentation was created for another audience. In that case, use it as a chance to bring the new listeners into a community of discussion. This makes them part of a larger whole rather than a castoff audience.

Wrong: I'm replacing X, who is much better at giving this talk.	Better: I'd like to thank X for providing these slides.

The form on the left implies that not only is the audience in for a bad talk but also that the superior speaker couldn't be bothered to talk with them. The approach on the right sends the message that the original (and often more accomplished) speaker really did care about the audience and provided their expertise to support the speaker.

Wrong: Linear; nonlinear	Better: One-for-one; accelerating
Wrong: Order of magnitude	Better: Ten times; one tenth

The phrases on the left are very meaningful to scientists but mean almost nothing outside science. Don't assume decision makers know them.

Wrong: What I want you to get out of this slide is . . .	Better: This slide demonstrates that . . .

The phrase on the left is often used as a substitute for a readable slide. Avoid the temptation to use complicated graphics that are unreadable or require too much time to comprehend (see chapter 9, "Designing Effective Visuals"). When you are forced to use a complicated graphic, such as when a short preparation period for a talk leaves you without the time or resources to redraw an important figure from a recent publication, draw the audience's attention to the result you find noteworthy by putting a yellow box around it is. This is a surprisingly effective way to pull one item out of a complex figure. Take the time to describe the axes of a graph or the meanings of the symbol types (which are often too small to read on a slide). If the graphic makes an important point in your talk, *honor the point* by making a slide that is legible and understandable to the audience.

AVOID SAYING THE WRONG THINGS TO AN INVESTOR

If you secure a meeting with an investor or a venture capital firm, then you have most likely passed the first hurdle by providing a concise and compelling executive summary that includes a clear answer to the question, "What problem am I solving?" Nancy Floyd says that typically scientists take the wrong approach at this important first meeting: "Too often we see technology looking for a problem to solve. We want to see the exact opposite—a solution that the entrepreneur has developed to solve a critical problem."[5]

She cautions scientists to show that they understand the competitive landscape:

> One of the big mistakes that scientists and management teams make coming into an initial meeting is to say, "I have no competition." It's classic. They overlook the fact that customers are doing something now to deal with the problem or issue, so a customer *not* changing what they are doing is competition for a new product. Doing nothing is always an option. In fact, it's the easy path forward for a customer. It's really hard for scientists who have spent their lives on something to understand this and to know the phrase "no competition" is one of the biggest red flags for an investor.

According to Nancy, another sad but true misstep for presenters making a pitch for venture funding in the United States is focusing the first one or two slides on how great your product will be for the planet or the environment. This depends on the investor, but it's generally a big red flag for the U.S. investment community. Doing good is great as a byproduct, but your stated goal needs to be to solve a real problem with a differentiated solution that you can sell profitably and hence build a successful company around. She also notes, "When I've listened to pitches in Europe it's just the opposite. In every single presentation I've seen, whether it's in France, Germany, Sweden, Norway, Spain, the first slide says something about helping the environment or the planet, or humanity. It is ok in Europe but not in the U.S."

Finally, technology is only one leg of the stool. Nancy points out that you have to show equal appreciation for customer value proposition, competition, and market opportunity—and that you should come in thinking in terms of profit and loss:

> Investors want to know that you are thinking about how to build a successful business, not a business you would love to own for 20 years but a business that is going to grow rapidly over five to seven years and generate the kinds of returns that investors expect. So, to come in and spend three quarters of the meeting on technology is a sure way *not* to get funded. You are thinking in technology terms when you need to show you are thinking in market terms.

AVOID ACRONYMS

So much of science has a language all its own (see "Jargon: The Specialized Words of Science" in chapter 5, "Why Scientists Communicate Poorly outside Their Field"). When we are in the classroom or with a group of scientists who share our expertise, acronyms are common knowledge. But that will almost never be the case when you are addressing decision makers.

Don't fall into the acronym trap. Using this shorthand creates distance between you and your listeners. If they can't translate the terms you are using fluently, you have managed to *distract* them from listening to you. If they have to ask what an acronym means, you have gone over their heads in a way that adds no value. Your ultimate goal is to make them feel smart. Help them understand your science so they can support your ask. Acronyms almost always get in the way of accomplishing this.

When you find you are tempted to use an acronym, see if you can take your message up a level or two to where broader concepts are conveyed and acronyms are no longer needed. If you aren't sure how much a decision maker knows about a technical area, assume it is not much.

DETERMINE THE RIGHT AMOUNT OF DETAIL

So how do you know how deeply you should delve into your topic? Circumstances help determine what your listener needs to hear. Are they being asked to make a decision? If so, what would they need to know to be sufficiently informed? Almost always, they need to understand how you plan to solve a problem, who will be doing the work, how long it will take, how success will be measured, and of course how much it will cost.

As soon as you can tell that your listener is tracking with your story, check in to see if they have an appetite for more information. This technique works very well in small settings when you are having a conversation, but you can also use it in a larger venue. Here's how.

Summarize the high-level information you just shared. If you were speaking about a company like ORIC, for example, you might say, "That gives you a sense of the focus of our work and our commitment to overcoming resistance in cancer. I've given you a high-level view. How many of you would you like more detail?" Ask for a show of hands. If more than half of your listeners indicate interest in taking a deeper dive, you need to try to gauge what kind of detail would be relevant. You might propose a choice by saying something like, "Would you like to know more about our multiple drug development strategies or about the mechanisms we are targeting that allow tumors to become resistant?" Ask for a show of hands again. The idea is to find a topic of shared interest quickly by suggesting a couple of options that you are prepared to discuss. If there isn't sufficient interest to warrant taking up everyone's time, offer to address specific questions during the Q&A or to chat after your talk with the listeners who want more.

A word to the wise from Rob Socolow of Princeton University: "Never underestimate a person's intelligence nor overestimate what that person knows."[6] Rob sees himself as a codebreaker. He told us:

> Each field has a private vocabulary that serves in part to keep other people out. Recently, I was introducing fuels and organic molecules in a class. Ahead of the class, I shared a slide that I had used many times in past years with my very astute grad student. He thought I didn't need to use it because everyone knew it. I told him that we can't assume everyone knows four very important but specialized words—*methane, ethane, propane* and *butane*. In class, I asked, "How many of you know what these four words mean?" Three hands went up. But you need to know these words. It's like conjugating the verb *to be*. I have to take the time to explain it. I tell them that learning a new field is largely about breaking codes. It's an essential first step on a path to communication.

Ernest Hemingway reminds us of the rigor required to make our thoughts clear and persuasive—even though he was talking about writing a novel: "Don't get discouraged because there's a lot of mechanical work to writing. There is, and you can't get out of it. I rewrote *A Farewell to Arms* at least fifty times. You've got to work it over. The first draft of anything is shit."[7]

The Words You Remove Add Power to the Words You Keep

This is the opposite of term paper mentality. It's no surprise that students leave school thinking that adding as much detail as possible is the right way to convey ideas. But when you start with a lot of good ideas and remove many of them, you are left with the truly outstanding ideas. When you want someone to remember what you say, default to the truly outstanding descriptions, not to a volume of lesser descriptions. Every time you remove something, think of it amplifying everything that remains.

Editing is one of the highest forms of being self-correcting. It can do wonders for improving your impact as a science champion. Edit yourself mercilessly. Remember the guideline in chapter 6, "The First Two Minutes," regarding the key message limit. Three is the magic number. It is the number of key messages that a listener can easily focus on and is likely to remember. With a decision maker, your goal is impact. The practice of editing your ideas is the foundation for getting to *yes*. While working on your message to convince a decision maker is less demanding than writing a classic novel, the same principles apply. Your phrases and constructions need to be clear and compelling. Your sentences must propel the listener to the next idea, not bog them down. The overall impact of a set of thoughts—shown on a slide or stated in a paragraph in a white paper—must bring your idea to life in a way that captivates the audience.

Steven Pinker—a Harvard professor, cognitive psychologist, linguist, and prolific science writer—provides a useful discussion in his book *The Sense of Style* of how we can make our writing simpler and clearer by avoiding muddy, confusing prose otherwise known as corporatese, legalese, academese, medicalese, bureaucratese, or officialese.[8] Citing Francis-Noël Thomas and Mark Turner, Pinker explores what he calls the "classic style of prose": "Its purpose is presentation; its motive, disinterested truth. Successful presentation consists of aligning language with truth, and the test of this alignment is clarity and simplicity."[9] Although Pinker's advice is directed at improving the way you write, his method is also applicable to speaking.

Pinker takes aim at the stodgy and even haughty style that academic writers often adopt. By contrast, he describes the classic style as "defined not by a set of techniques but rather by an attitude toward writing itself.

What is most fundamental to that attitude is the stand that the writer knows something before he sets out to write, and that his purpose is to articulate what he knows to a reader."[10]

The writer's goal is to get the audience to focus on an issue or fact that you want them to absorb. Pinker explains, "I have tried to call your attention to many of the writerly habits that result in soggy prose: metadiscourse, signposting, hedging, apologizing, professional narcissism, clichés, mixed metaphors, metaconcepts, zombie nouns, and unnecessary passives. Writers who want to invigorate their prose could try to memorize that list of don'ts. But it's better to keep in mind the guiding metaphor of classic style: a writer, in conversation with a reader, directs the reader's gaze to something in the world."[11]

Done well, presenting and writing reflect a clarity of thought in both preparation and delivery.

SHOULD YOU GO PASSIVE? THE PASSIVE/ACTIVE VOICE CHALLENGE

Steven Pinker's book on style brings up another very important discussion on the topic of scientists talking to people outside their field. Using the passive voice is often frowned upon because it is boring and impersonal. And yet so many scientific talks and papers use this voice. You might find yourself thinking it is unprofessional to use "I" in conversation, let alone in print. How do you decide when to use the passive voice and when to use the active voice? Are there rules scientists can apply? It turns out there are.

Too often, people use the passive voice to avoid direct responsibility. The politician's "mistakes were made" is an egregious example,[12] but scientists have become addicted to the passive voice because they think it incorporates the entire body of work that has been done in their field. When you say (When it is said?), "The work was conducted in our laboratory," you feel like you have given credit to your whole team, all the administrators, and the guys who properly installed the plumbing. But you have cheated the decision maker of the thing they would find useful to know—who played the major role?

This is where Pinker points out that the choice between the passive voice and the active voice allows the writer to direct attention: it lets the listener know how they should be viewing (in their minds) the scene you are setting for them. These are Pinker's key ideas from his excellent extended discussion of the passive voice:

> The passive voice . . . has several uses in English. One of them . . . is indispensable to classic style: the passive allows the writer to direct

the reader's gaze, like a cinematographer choosing the best camera angle. Often a writer needs to steer the reader's attention *away* from the agent of an action. The passive allows him to do so because the agent can be left unmentioned, which is impossible in the active voice. You can say *Pooh ate the honey* (active voice, actor mentioned), *The honey was eaten by Pooh* (passive voice, actor mentioned), or *The honey was eaten* (passive voice, actor unmentioned)—but not *Ate the honey* (active voice, actor unmentioned). . . .

Even when both the actor and the target of an action are visible in the scene, the choice of the active or passive voice allows the writer to keep the reader focused on one of those characters before pointing out an interesting fact involving that character. That's because the reader's attention usually starts out on the entity named by the subject of the sentence.[13]

Scientists can take this rule and apply it to everyday presentations and conversations. Consider whom you want the listener to think about in the sentence you are about to say. And remember that decision makers are very focused on who the most effective person is in a field. If you want them to know it is you, say "I." However, it is rare that you can unambiguously say that, so rather than defaulting to the passive, think about editing the nature of the accomplishment you are describing. "I pursued this topic for five years" places no burden on the rest of your field to be good or bad—it describes what you did. "I came to the conclusion that . . . " is a perfectly honest way to tell the decision maker what is in your head. If, on the other hand, you want to convey a sense of what a broader set of minds is thinking, it is perfectly all right to say, "In my field, it is widely accepted that . . . " But keep Pinker's camera in mind:

> The problem with the passives that bog down bureaucratic and academic prose is that they are not selected with these purposes in mind. They are symptoms of absent-mindedness in a writer who has forgotten that he should be staging an event for the reader. *He* knows how the story turned out, so he just describes the outcome (something was done). But the reader, with no agent in sight, has no way to visualize the event being moved forward by its instigator. She is forced to imagine an effect without a cause, which is as hard to visualize as Lewis Carroll's grin without a cat.[14]

9

Designing Effective Visuals

Anyone can look at a graphic and immediately tell a good one from a bad one, but the question is, why is one better? People subconsciously notice the use of contrast, alignment, and relationship. These are simple mechanisms you can use. I want the viewer to immediately see what is most important and what is of secondary importance. I don't want to distract people with a whole jumble of stuff. If the point of the slide is one figure, then I make it the whole slide.

—Joshua White, engineer and geophysicist

Since almost all science communication involves visuals, you have a terrific opportunity to expand your listeners' understanding and make your message much more memorable by tying it to good graphics. Once you have masterminded your message, your next investment in effective communication is to develop an equally well-crafted set of visuals.

All too often, scientists try to pack more into a single visual than the viewer can possibly absorb—and so they absorb almost nothing. In presentations (as opposed to publications), it is better to let your visuals focus on a single salient item that you want the listener to understand or be impressed by.

After the space shuttle *Challenger* was destroyed in a fiery explosion in 1986, a Congressional inquiry was launched into what went wrong. The investigation found many instances where scientists thought there might be a problem before the disaster occurred and an almost equal number of examples of how those concerns were confused, downgraded, and muddied by bad slides and inadequate descriptions. There *was* a problem, and many people recognized it before the accident, but amid the thousands of items that had to be considered before a shuttle launch, it got lost in the noise.

There turned out to be a long list of concerns that never made it up the chain of command. The famous physicist Richard Feynman was asked to serve on the inquiry board, despite having no particular experience in rocketry. He saw all of this confusing evidence and realized that the key was to extract the essence—to focus attention on the aspect of the failure that appeared to matter most. Feynman had seen a huge pile of evidence that pointed to a failure of the O-rings in the boosters. And he had seen testimony from engineers who said that those O-rings might have been inappropriate at the temperatures in Cape Canaveral that winter morning.

Feynman chose a very simple and compelling visual to explain his understanding. He did not make a graph of O-ring flexibility versus temperature. He did not lead the members of Congress he was facing through a detailed explanation of how O-rings work in engineered devices. He simply took a small O-ring made from the same material as the large ones in the booster rockets, and dunked it in his water glass—conveniently ice water—at the Congressional hearing. He had to double the thin, flexible O-ring over in order to force it into his glass—and when he removed it, the shape remained the same. It did not spring back to the original form the committee members had clearly seen a minute before. Feynman said, "I believe that has some significance for our problem."[1]

The before-and-after nature of Feynman's graphic demonstration made the problem clear. He extracted the essence in iconic terms that the viewer understood from their experience with ice and cold. Feynman had delved deeply into the physics of the problem, but he spared his listeners those details.

This demonstration launched an entire cottage industry of evaluating the exact details of how the failure occurred and how the concerns about temperature failure did not come forward before the accident. But the simple, good graphic got the salient message across to the members of Congress that day. And Feynman did not have to support that essence with temperature calculations for that fateful Florida morning—he left that to others and to extensive volumes of reports that would never be read by members of Congress. He gave the decision makers the visual they needed.

CHARACTERISTICS OF GOOD GRAPHICS

What constitutes good graphics in your discussions? Human brains are pattern recognition engines, and a good graphic engages the brain extremely effectively. Graphics allow color, shape, and relative position to amplify static numbers. Good graphics capitalize on all of the components to make the more complicated parts of your message understandable. They do this in the context of things your brain already understands.

Unfortunately, graphics are too often used inappropriately to brag about how much work was done, such as when a graph is filled with incomprehensible data points. When speakers make this kind of graphic, their goal is usually to support the assertion that they are, in fact, smart enough or hard-working enough for the decision maker to take seriously. It rarely works. The only situations where those goals are appropriate would be your thesis defense or some oral examination—for which you are unlikely to have prepared graphics anyway! Save the extensive details for your publications.

Scientists spend an enormous amount of time preparing graphics for publication in journals. If you use these—as is—in your talks, you are abusing the audience. They can pick up your paper if they want to spend hours pouring over the nuances of detailed charts. (Your professional society and journals will have awesome examples of how to prepare great graphics for publication.) "But," you exclaim, "I spent days on those graphics, and I don't have time to create new ones!" Fair enough. The key here is impact through simplicity, through extracting the essence (see "Capturing the Essence Visually" in chapter 3, "Extracting the Essence"). Let your listeners benefit from your learning by giving them an explanatory summary.

You disrespect your listeners' time if you deluge them with details that they don't understand. The complexity of your graphics must align with how well your audience already understands the topic and how much time you are willing to take to explain it. For the champion talk we are focused on in this book, the assumption is that much of your audience does not have a good grasp of your science and that you want to get important concepts across quickly. Your approach to a department seminar can be much more detailed.

Roger thinks about good graphics in the context of a whiteboard drawing. When you are in your office describing a new concept to a colleague, you don't try to draw a bunch of data in detail. That would take more time than you typically have for that kind of conversation. You emphasize general trends and relationships. You want your colleague to see the key relationships as you see them. Keep this view in mind as you create graphics for presentations to decision makers.

Many presentations are mostly words and dense graphics. It has become commonplace to see three or four graphs on one slide. This is often a symptom of the speaker not really thinking through what the important point should be, so they just put everything on. Slides like that are just lazy. Don't become that speaker. Take the time to prepare an appropriate number of high-quality, *single-topic* graphics that support your key messages. Think about what matters. Clear, understandable graphics take time to create and often require some professional help.

If you are giving an informal lunchtime talk to your work group, you might only require one new graphic with details of your latest result. But if you are proposing new funding or providing a plenary presentation for a conference, *every single slide* deserves your careful attention. Don't look like you wrote the talk on the bus on the way over. Respect the audience, and you will enhance your reception.

GRAPHIC TYPES: DISCOVERY, EXPLANATORY, AND VISUAL INTEREST

There are three basic types of graphics used in a presentation about science: discovery, explanatory, and visual interest.

Discovery Graphics

Discovery graphics are often simple reproductions of the computer plots used to reach important conclusions. Because scientists tend to have lots of these stacked up on their computers, they often rely on them for presentations. Resist the temptation, because these are often of low value in a presentation (with a key exception that is discussed below).

A discovery graphic generally requires the audience to have the same background as you in order to fully appreciate its impact. And while that might be appropriate in seminars for your peers, this type of graphic should be used very deliberately and sparingly with decision makers. The one valuable and appropriate place to use a discovery graphic is to establish with your audience, just once, that you really understand your science and that they can trust you when you give them summary descriptions later. Since the goal is not to lead them through each of your discoveries but to establish your credentials *to discover* things, you usually don't need more than one discovery visual per talk. And that is good, because these graphics are time-consuming to prepare and narrate properly.

The Three Types of Science Graphics

Discovery, explanatory, and visual interest graphics have different purposes.

- *Discovery* graphics are often simple reproductions of the computer plots used to reach important conclusions. Use these to establish your credentials to discover things.
- *Explanatory* graphics are akin to what scientists draw on their whiteboard during a casual conversation in their office. They show relationships and trends and help simplify complex topics. This is the most important type in a presentation to a decision maker.
- *Visual interest* graphics are often photographs, and they help the viewer place your material in a visual context.

Explanatory Graphics

Explanatory graphics are what scientists draw during a brief, casual conversation in their office. They show relationships and trends and help bring simplicity to a complex topic. They are the most valuable graphics in a presenter's portfolio and should be in common use to amplify the impact of a dialogue.

Visual Interest Graphics

The third graphic type is *visual interest*. This is the graphic that is fun to look at. You will see many examples of these in books on business presentations like the one in which a cheery-looking poison dart frog had a text bubble highlighting a key point. This might be a way to get your audience to remember that point but should be used very sparingly in science because it can come across as flippant. More useful examples of visual interest graphics in science might be a photo of a field site, a close-up of new material, or a photo of your experiment splattered on the wall after you miscalculated its enthalpy of reaction. These graphics help the viewer place your material in a visual context, and humans are visual learners.

Most of the graphics in a good presentation are explanatory and visual interest. They should always be more prevalent than word slides. We will return to the best ways to use word slides at the end of the chapter—sometimes you need them, but never start with them!

CRAFTING DISCOVERY GRAPHICS

Let's begin our examples of great presentation graphics with the one you have made most often but that we bet has been among the *least* impactful in your portfolio. A discovery graphic demonstrates that you understand the way data is collected in your field. It is an important deep dive into your scientific field that, when done correctly, makes it possible for you to make a lot of other points in summary form. It is not important to discuss each individual point, which will put your audience to sleep no matter how much they love your topic. Rather, you should use it to tell a meaningful *story* about your discovery of what the data means and, along the way, make it clear that you are a competent scientist.

A slide with data on it is an exceedingly difficult thing to make understandable to the viewer. You are very familiar with it, but to the audience, it is just a blast from the fire hose. Build the content slowly, and explain why this visual matters. A great way to create a context for your audience is to narrate your discovery graphic using Randy Olson's And-But-Therefore approach described in chapter 7, "Crafting Key Messages and Narratives."

Rather than saying, "Here is all the data I collected over the last two years," it's much more interesting to tell the story this way: "When I first started out, I thought this way about the topic, *and* I collected some data reflecting that thought, *but* you can see that this data is not entirely consistent with that thought, *therefore* I developed this new understanding that helps me explain much more of what matters in this problem"—all in one graphic.

Let's take a look at a wonderful example from William Ellsworth, a seismologist at Stanford University and the U.S. Geological Survey (see figure 7). We have redrawn his figure as you might for a presentation. This single graphic (and of course the analysis it embodied) changed attitudes about underground injection of wastewater in the Central United States. It clearly demonstrates something that geologists knew to be true but had previously been unable to quantify in a way that would enable regulators and policymakers to act decisively. It was well known that injecting water deep underground could trigger earthquakes, but how many earthquakes? Was this a minor nuisance or a major hazard?

Ellsworth got his discovery graphic off to a good start by representing the state of understanding before the discovery. He did that with a

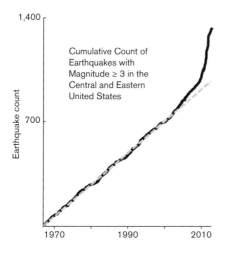

1,400

Cumulative Count of
Earthquakes with
Magnitude ≥ 3 in the
Central and Eastern
United States

Earthquake count

700

1970   1990   2010

FIGURE 7. Presentation-quality version of William Ellsworth's famous earthquake plot (2013). Original figure from William L. Ellsworth, "Injection-Induced Earthquakes," *Science* 341, no. 1225942 (2013), DOI: 10.1126/science.1225942). Redrawn by the authors as would be appropriate for a presentation slide.

trend line that reflects the expected rate of earthquakes. This is the dotted line in figure 7, Earthquake Frequency in the Central and Eastern United States, from Ellsworth's 2013 paper in *Science*.[2] We don't expect the overall frequency of earthquakes to change over time. Each year, we anticipate roughly the same number of earthquakes to occur given we are considering a large enough footprint. The total number of earthquakes we experience in that footprint area adds up year after year. The dashed straight line graphically shows that expectation. Ellsworth could have alternatively said that the number per year (rate) is constant, but that generates a less interesting flat line. By using the *cumulative* number of earthquakes, he came closer to our human experience.

Along with the expected rate, Ellsworth includes the solid line, which is actually composed of hundreds of dots representing individual earthquakes—*and* this is what the data looks like for the Central and Eastern United States over the past forty-five years, in Olson's phrasing. The *but* practically bites you without any additional explanation: "*But* in the last few years, there has been a dramatic increase in the cumulative number of earthquakes." If you were explaining this to decision makers, they would be riveted to your next statement. They would want to know what you discovered. Why is the number of earthquakes increasing? Ellsworth points to the recent increase in disposal of water from hydraulic fracturing operations into deep injection wells, particularly

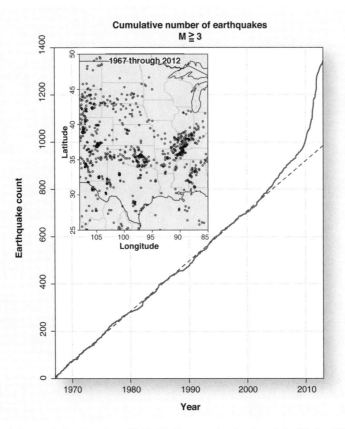

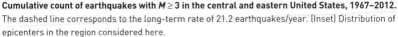

Cumulative count of earthquakes with *M* ≥ 3 in the central and eastern United States, 1967–2012. The dashed line corresponds to the long-term rate of 21.2 earthquakes/year. (Inset) Distribution of epicenters in the region considered here.

FIGURE 8. Original version of Ellsworth's figure, as published in *Science,* along with the caption. Figure from Ellsworth, "Injection-Induced Earthquakes," 1225942–2.

water used in hydraulic fracturing of shale: "*Therefore* we considered the possibility that this increase in earthquakes is associated with disposal wells, and indirectly then associated with the increase in hydraulic fracturing activity." The remainder of Ellsworth's paper is an evaluation of this hypothesis.

Figure 8 shows the more complicated version of this figure as it appeared originally in the *Science* article. Remember, archival figures can have a different structure than your presentation visuals. The added map with earthquake locations has an enormous amount of information. The grid allows the viewer to examine exactly which years had slightly higher or lower rates. The labels are large and in a clear font. There are a minimum number of tick marks.

In the presentation version (figure 7), there is no label for year—that information can be readily assumed from the numbers. The expected trend and observed data are obvious and require no added labeling. The grid is dropped since the clear vertical association with the year is obvious, and more importantly, in a talk we do not expect anyone to get out their ruler to see which year corresponds to changes in the number of earthquakes. If they want to know, they can ask you! Leaving these details out lets your audience focus on the change in the number of earthquakes that appears *approximately* at the time hydraulic fracturing increases.

The shaded border of figure 8 helps the viewer's eye see the figure on the page of the journal. We have removed it in the presentation version of figure 7 since each slide usually has a figure and you don't need to draw attention to it. Notably, there is no legend in either version. A legend forces the audience to shift their attention back and forth across the slide, and they lose track either of what things mean or of what you are saying while they try to process with both their ears and eyes. Make it simple for them.

Edward Tufte exposes simplifying approaches like these in his remarkable books on scientific graphics.[3] His *The Visual Display of Quantitative Information* was the first book to provide invaluable guidance to scientists on how to make their figures more useful and understandable. Tufte's single overriding principle is that you should always use minimum *non-data* ink. Any ink on the chart that is not explicitly data should be there only so viewers will know what the data is showing. Put another way, every line or word on a slide is a piece of information that your listener has to observe and understand. Make it easy for them by only putting things there that matter. Many graphics programs make themselves seem more valuable by overloading your initial product with these grids, tick marks, and side legends. Fortunately, there is a delete key. Use it.

One thing that we don't want to skimp on is the actual data. In both figures 7 and 8, the data are all explicitly shown. There is no attempt to create a cloud around it or to describe any error in the data. By showing all the data, the audience gets a very good idea of what the difference might be from year to year—and it is very small. The final years' divergence from the trend line is obviously outside any margin of error or uncertainty. And that makes it easy for you to discuss the uncertainty in this slide—that is to say, you *don't* discuss it. The audience gets it completely from the graphic, and you can spend your time on the *therefore,* which is what matters.

Figure 7 is a very simple discovery graphic, and yours may be more complicated. Let's talk about some of the ways you can enable your audience to quickly understand what you did and why it demonstrates your mastery of the topic. If you have multiple classes of data, organize them with color *and* shape. Not everyone sees color the same way, but shapes are often too small to see on a projected image. Using both helps ensure that it will be understandable to everyone. If you understand the sources of error or uncertainty, show a representative error bar—but don't clutter the chart with them. You simply want people to have a sense of the degree to which the data should fit the expected behavior you have described.

Be extremely cautious of the technique that Jay Davis calls "pulling the veil." In this approach, you expose parts of the data as you build up a complex slide from a set of simpler components that add to each other. Why do you want to be cautious of this? First, absent a story, it is insulting to the audience. It seems like you don't trust them to understand the whole picture. The people you most likely want to understand this data are the members of the audience most familiar with the topic. Give it to them all at once so they can absorb it with their eyes while you talk.

Second, staging a slide with multiple reveals encourages you to put too much on it, thinking that your artful buildup makes it more understandable. Far better to have multiple slides if you need to cover all that material. If there *are* multiple facets that you need to expose separately, a visually compelling method is to show all the data in grey and then color individual sets as you describe them.

Why did we say "absent a story"? If there are multiple stages to your and-but-therefore story, building it up in that order can make a compelling human-interest story. "Here is how I proceeded along this journey, let me show you the sights I saw." This is how Roger tells the story of the evolution of U.S. electricity generation that we saw in three slides at the end of chapter 5, "Why Scientists Communicate Poorly outside Their Field." This approach can make a story interesting for everyone, no matter their familiarity with the material. That's not pulling the veil, it's telling a great story. With science pictures.

Finally, take your time. The discovery slide has a lot of weight since you are mainly using it to establish your bona fides. Use precise language in describing it. Don't ever use the phrase "playing with the data" or other language that downgrades how important this work is. It's acceptable to be shy, and it is great to be humble and acknowledge how others helped you accomplish these scientific feats. But most of all, take this slide very seriously. Watch the audience to make sure they get it, and be prepared to answer questions.

EXPLANATORY GRAPHICS THAT GET THE MESSAGE ACROSS

In explanatory slides it is usually unnecessary to show your data points or error bars explicitly. You are showing relationships and trends. You are teaching the listener about the structure of this aspect of science. You are back at the whiteboard, but with lots of time to prepare.

Each graphic should have a clear message—a reason you are showing it. Think about that message first, and craft around it. Don't fall into the trap of using a graphic you already have, or one you borrowed from a friend, for a key point. Science deserves to be done carefully. Remember the recommendations about crafting key message content? Here's your chance to bring those messages to life visually. If you do this well, you will make memorable images that underscore the most important points of your talk.

What is the key failing of explanatory graphics? They try to explain too much. A presentation requires the viewer to understand the graphic, and your key message about it, in less than a minute. Are you tempted to tell two stories in a single graphic? Separate them. This can be a clever way to help the listener follow an evolving set of conclusions. By first showing an explanatory graphic (and taking the time to be sure the audience understands it), you can later show a second graphic formulated in a similar manner but with key differences. Because the human brain is so good at pattern matching, most audiences will follow this progression readily.

Rob Socolow and Steve Pacala's slides on climate change abatement (the wedge diagrams) that we saw in chapter 3, "Extracting the Essence," are a great example of explanatory graphics that adhere to Tufte's principles. There is one topic per slide. The graphics are simple and clear. Rob uses enough words to make the visuals understandable without his voiceover, but no more. The axes are simple. But most important, the wedge diagram is an outstanding example of the whiteboard ethic. Rob and Steve's key concept was that no new exotic technologies were required to solve the climate problem, just constant application of technologies that exist, at reasonable levels. These diagrams elegantly expose that thought process.

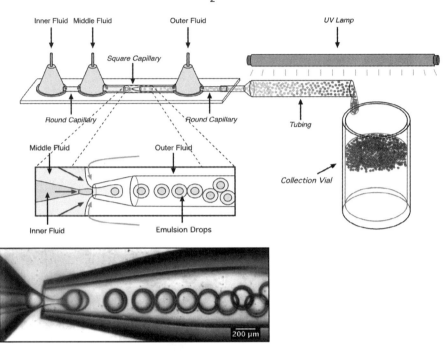

FIGURE 9. A slide explaining how Josh Stolaroff's encapsulated solvent approach works. Figure by Joshuah Stolaroff, photograph by Congwang Ye.

Joshuah Stolaroff of Lawrence Livermore National Laboratory is a master at the process of turning images into figures that make effective explanatory graphics. He and Roger worked together on the creation of encapsulated solvents, in which very small beads are formed with liquid cores and permeable polymer shells. These capsules absorb CO_2 so that it can be safely separated and stored away from the atmosphere. Figure 9, made by Josh and his colleague Congwang Ye, demonstrates how these capsules are made.

The image at the bottom is a video embedded in their presentation. Just above the video we see a process diagram depicting the elements in the image. At the top of the diagram there is a simplified drawing of the device used to make the capsules. The voiceover describes how all of these fit together, but you can easily get the gist of it without additional help.

Videos are a brilliant way to convey a wealth of information about complicated topics—if you keep them simple. When Josh shows the slide above, the video of the capsules being produced is mesmerizing, very much like a lava lamp. The audience has trouble looking away from it to check out the rest of the image, so he just runs the movie once through and then lets it stop, after about fifteen seconds.

Normally we recommend looping videos continuously so that the viewer can have a couple of chances to understand what they are seeing while you talk over it. Visual images are easily understood once you have a framing knowledge of what they are—this can take one or two loops to establish. Presenters, realizing the difficulty of understanding videos on the first viewing, often launch into a lengthy explanation— "What you are about to see is . . ."—before starting the movie. Don't fall into this tar pit. Launch the movie, and then describe *what they are seeing*. Better yet, set all your videos to start automatically when the slide comes up. Then you are not tempted to explain first. The movie is always more interesting than your words! Short fifteen- or twenty-second clips of your device running, a landslide coming down off a mountain, or a crash test are absolutely the most impressive thing your audience can see in your presentation. Properly framed with explanatory information, a short video is the difference between ho-hum and stunning.

Speaking of ho-hum, let's consider one of the most common data displays, the Excel graph. If you ever find yourself tempted to use an Excel graph *as is* in one of your presentations, your self-awareness alarm bells should be blaring. And you better do some quick self-correcting, because if the information is important, it's worth transmitting well. Take a few minutes to improve it.

It is not difficult to clean up the graphs from plotting programs to make them usable in presentations. Figure 10 shows a graph Roger made of the daily electricity use in California, which can be downloaded in table form from the California Independent System Operator (the wholesale organizer of electricity in the state). The graph is accurate, but very hard to read even in print. In a projected slide, viewers would get little more than the overall shape. The order of the energy sources is not particularly aligned with any message Roger wants to convey about this data. Viewers would be hard pressed to tell which patterns correspond to which power source, although at least Excel does present them in the same order as in the legend. The grid lines serve no purpose, nor do most of the ticks and labels. Time for the delete key.

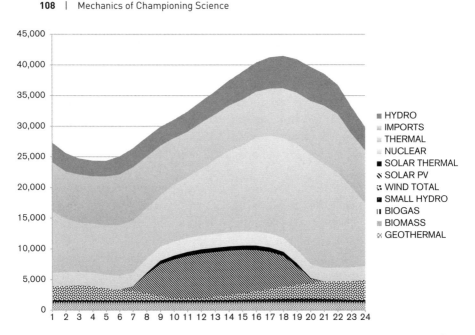

FIGURE 10. An Excel version of a graph showing one day's electricity use in California. Figure by Roger Aines. Data source: "Daily Renewables Output Data for 07/14/16," Renewables and Emissions Reports, California ISO, July 14, 2016, http://content.caiso.com/green/renewrpt/20160714_DailyRenewablesWatch.txt.

Figure 11 shows the result of Roger investing about thirty minutes to make it more legible. He reordered the data so that the renewable sources of power are together on top. He filled the space from left to right and put the labels inside the graph. He ordered the labels so that they are easy to associate with a graph element. The font size is large enough to read from the back of any room. He got rid of superfluous labels—we all know which hours come between 6 a.m. and 12 p.m.—and used minimal y-axis labeling, because it is the relative size of the elements that is important. He used a black background to make the colors pop—this is good practice in large rooms that are dark. He moved the title inside the graph to the otherwise wasted space in the upper left. The data fills the slide area, without boring white (well, black in this case) space around it.

To emphasize his message, Roger added the arrow labeled "renewables." Now a quick glance gives you an instant sense of the story and a piece of important information about California's summer electricity supply.

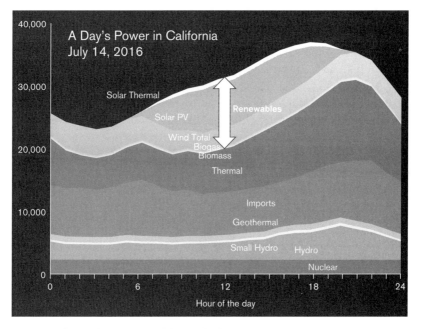

FIGURE 11. The revised version of the graph shown in figure 10, after being cleaned up to reinforce the key message. Figure by Roger Aines. Data source: "Daily Renewables Output Data for 07/14/16," Renewables and Emissions Reports, California ISO, July 14, 2016, http://content.caiso.com/green/renewrpt/20160714_DailyRenewablesWatch.txt.

If this were a discovery graphic—if Roger had created an instrument that made this measurement for the first time—he would have wanted to keep all the details. But this is an explanatory graphic, and the viewer just needs to see the shapes and have an idea which power sources they represent. The means that were used to obtain the data are not important, and data is almost a distraction in an explanatory slide. Include it in the most representative way possible, but without detail. Let the listener spend their brainpower on what the data *means*.

Don't try to condense your talk by putting multiple graphics on one slide. If all of the graphs or figures on a slide are important, then give them each a slide and *advance the slides at an appropriate pace* to keep the presentation on schedule. Full-size figures are much easier for an audience to absorb, so they don't have to be on the screen as long. Well-designed figures, one per slide, actually save time. Don't let the *number of slides* be your determinant—think about the *number of points* you want to make.

VISUAL INTEREST: THE EYE CANDY OF YOUR PRESENTATION

Just as Tufte taught us how to display data, Felice Frankel showed us how to take and use photographs in her classic book *Envisioning Science* and expanded the lesson into using photographs as the core of great figures in *Visual Strategies* (cowritten with Angela DePace).[4] Frankel's and Tufte's books form the nucleus of any good library on scientific graphics—although you may want to visit your library to read them, as the price of any high-quality book on graphics can be daunting. Spending time with these books will show you how truly transformational great scientific graphics can be.

Frankel, a member of the MIT research faculty, has spent her career teaching scientists to make better images, photographs, and visual representations of their science. She criticizes scientists for trying to overload their images with information, reflecting Bill Young's "I did all this work, and I'm going to show it to you!" mindset. Of course, when you are so familiar with a topic, complex images are perfectly understandable. For your audience, according to Frankel, "It doesn't work that way."[5] Editing is not just an activity for words. As images become more and more central to science, we have to pay attention to their information density as well. Here though, we can take advantage of the fact that the human brain is highly optimized to rapidly assess image information.

Frankel is a proponent of using images of similar objects, or *visual metaphors*, to help nonexperts understand technical topics. In her book *Envisioning Science*, she gives the example of using an image of a piano roll to describe how a DVD works. Roger was thinking very much of her approach when he first used the image of a locomotive (figure 12) to describe how large a hydrofracturing pump is in a modern oil drilling operation. In his talk, he explains that the pump uses the same diesel engine as the locomotive—and that in the oil field, *twenty* of those pumps are ganged together to perform the hydrofracturing. The listeners don't need to know the details of exactly how the pump works for them to be very impressed by the enormous power it represents.

Frankel discusses the process of how an *image* becomes a *figure*. As scientists are preparing photographs to use in explaining science, she counsels them to remember that the viewer can be overwhelmed by all the detail in a photograph—detail that you tend to automatically edit from your view (remember how surprised you were that power lines suddenly appeared in your photograph of a beautiful vista when you hadn't noticed them in person?). Once again, extracting the essence and

FIGURE 12. A visual metaphor—the engine in this locomotive is one twentieth of a modern oil field hydrofracturing pump. Photograph by Terry Cantrell.

avoiding clutter are key. Keep the number of elements in your photographs to a minimum, and provide labels, highlighting, or a simple process diagram to help the audience understand what they are seeing.

One last thing about when a photo is used as a discovery slide: it should be an image of *what* you discovered and tell a story about *how* you discovered it. It should never be an image of your ragtag experimental apparatus with a voiceover of what you found by using that apparatus. That may seem easier, but have you ever seen a picture of an experimental apparatus that really impressed you? No, they all look like Tinkertoys. Spare the viewer, unless it becomes an interesting element of your story, such as, "Here is the red color that I kept observing, not knowing that a year later I would realize that the new compound had been in my flask all along." And if you do that, zoom in on the flask.

Photo slides introduce visual interest into a presentation. Felice Frankel and Garr Reynolds call this the *picture superiority effect.* Looking at text and tables requires significantly more effort than looking at pictures—it is a function of how our brains evolved. When you put a clear, interesting picture into your presentation, it gives the audience a break from trying to understand difficult text or mathematical forms and gives them something that they can immediately absorb. They will also remember it much better than any other type of information. The slide of the locomotive (figure 12) is an example of this—it conveys a distinct message and does so very rapidly.

FIGURE 13. A slide of Joshuah Stolaroff's encapsulated solvent IL-SiTRIS capsules (see figure 9). A full-bleed photo slide like this provides a great visual break for the audience. Viewers will absorb the information rapidly because our brains process images faster than words. Figure by Joshuah Stolaroff, photographs by Congwang Ye.

Josh Stolaroff followed Reynolds's and Frankel's advice when creating the visual showing his completed capsules in figure 13. The photo is beautiful—it's enjoyable to look at it, and the reflection provides an additional angle to highlight the translucence of the capsules. To the right is a decidedly less beautiful image of the experimental apparatus with the capsules loaded into it—it conveys information about the way the experiment is conducted, as well as the size of all the equipment. This is a classic example of punching up a visual interest slide with a little more content, satisfying multiple goals.

Note that there is no frame around the image—it is presented "full bleed," which means reaching all the way to the edge of the projection space. Really, *all the way to the edge!* And covering up any logos or lines that might be in your organization's slide template. Make the image the only thing that the viewer sees, *no* white border. This makes the image completely take over the viewer's attention. It is always the best way to handle photographs, but it is done so rarely that it can really impress a science audience.

SOURCES FOR GOOD PRESENTATION DESIGN

Where are the Tuftes and Frankels in the area of explanatory and visual interest graphics? Although much can be found on these topics in their excellent books, on balance they tilt toward making excellent descriptions, rather than explanations, of science. For explanatory graphics we can find a lot of help in books fundamentally oriented toward business. These have to be read with a little fortitude. You will find yourself skipping sections on marketing pitches and product placements, but it is very much like the instruction book for your new mass spectrometer. The majority of those machines are sold to businesses that conduct very different work than do you, but the instructions are still valid. Read the business-oriented books with the same detachment. Look for the pearls of wisdom that come from people who make their living by getting ideas into the heads of decision makers, and don't get distracted by the business fluffery.

Unfortunately, we still think that the great majority of business books on presentations make better kindling for a campfire than instruction for scientists. However, two authors definitely deserve your attention—Garr Reynolds and Nancy Duarte. We first met Reynolds in chapter 3, "Extracting the Essence," where his ideas of simplicity in message development helped us see that first you must assess what is important—and what is not. His *Presentation Zen, Simple Ideas on Presentation Design and Delivery* is our favorite book on simple, elegant presentations.[6] It's a little hard to start due to its focus on Japanese examples and metaphors, but once you realize that these are terrific ways to talk about presentation issues, you'll find that the book is filled with gem after gem. It holds a vaunted position on Roger's desk, along with Strunk and White's *The Elements of Style,* as a book that he picks up every year or two to reread just to make sure he keeps all the great advice at the top of his mind.

Reynolds's advice on good graphics parallels his advice on messages: "In the world of slide presentations, you do not always need to visually spell everything out. You do not need to pound every detail into the head of each member of your audience either visually or verbally. Instead, the combination of your words, along with the visual images you project, should motivate the viewer and arouse his imagination, helping him to empathize with your idea and visualize it beyond what is visible in the ephemeral PowerPoint slide before him."[7]

In other words, use graphics that *lead* your audience to the ideas rather than jamming the ideas down their throats—which in reality won't work anyway because viewers rarely get it when you use complex and obtuse graphics. Stay simple, and let the audience learn something rather than get drowned by something.

Reynolds's metaphorical comparison of presentations and Zen is focused on enabling the audience to see the material that is most important. He likens a presentation to a garden: "A Zen garden is also a lesson in simplicity. Open space with ornamentation, a few rocks carefully selected and placed, raked gravel. Beautiful. Simple. The Zen garden is very different from gardens in the West that are absolutely filled with beauty, so much beauty, in fact, that we miss much of it. Presentations are a bit like this. Sometimes we're presented with so much visual and auditory stimulation in such a short time that we end up understanding very little, and remembering even less."[8]

Following Reynolds's advice on good graphics results in giving the viewer room to see themselves in your idea. What do we mean by that? Everyone has different experience and slightly different goals. If you make a graphic extremely detailed, you force them all to see it through the lens of your experience and your goals.

While this may seem like your intent, remember that you are trying to influence people. Engaging their previous experience is a powerful tool, whether you do it with visuals or the way you tell a story. Rather than telling your audience exactly what you think the future should look like, you have the opportunity to engage them as partners in creating a shared view.

Roger learned this lesson when he was developing technology to clean contaminated groundwater using steam and electric heating of the soil. His team made several very detailed movies of the process—which ended up having no impact at all on audiences despite having consumed a notable part of the budget! When he checked in with multiple viewers, he found that people already had a visual image of how the underground process worked based on their understanding of things like washing machines. When he showed the video, it either confirmed that understanding (and wasted their time) or conflicted with that understanding (and created a divisive point in the presentation). The issue that he was trying to get across was the *value* of using these methods to clean up contaminated soil. The actual *mechanism* by which they worked was not as important and was better explained with words that allowed the audience to engage their own mental images.

Reading a book like *Presentation Zen* will also give you tips on how to make your slides look really good. This might seem slightly unscientific, but in fact it is highly respectful of your audience's need to understand your material quickly. Making the images easy to absorb gives viewers more time to think about your message and more time to think about how they could contribute toward your goals. Some of the key skills Reynolds discusses are using white space to highlight important material and keeping material on each slide in balance.

One of the most impressive scientific champion talks of all time was Al Gore's climate lecture series, which was made into the movie *An Inconvenient Truth*. Gore's slides were created by Nancy Duarte, the CEO of Duarte Design in San Jose, California. (Many of the slides can be seen on the company's website, in the portfolio section.)[9] Nancy's book *Slide:ology The Art and Science of Creating Great Presentations* is an amazing deep dive into how to get the most impact from every idea. If you are like us, this book will confirm, page after page, how little you know about graphic design—it is like being shown calculus when all you know is arithmetic. The power of the highly evolved graphic design that Nancy uses is as elegant, as useful, as difficult to become good at, and as hard as calculus. Most people will never approach her level of skill. To attempt to do so would seriously distract us from the science we are trying to champion. However, there are several wonderful concepts to learn from a few hours with *Slide:ology*. We encourage you to spend a rainy Sunday afternoon with it sometime to get a sense of what sorts of techniques you, or your graphic arts support, could be using.

Color is important in slides—it makes images visually attractive and can color-code key concepts. If you are going to use color as a substantive part of your presentation—for instance, in a series of graphics or as consistent text highlights that refer to graphic elements—you may want to pick a different color scheme than the default ones in your slide software. Nancy discusses color choice in detail and gives lots of nuanced examples of how to make good choices. The fundamental element is what's known as Isaac Newton's color wheel, which you see when you choose colors from the palette in your computer. It is just one of many ways to represent the colors you can see. It is convenient. Nancy will teach you how to think about the color wheel as a tool—keeping colors concordant, making highlights effective, and blending colors from your images with your text.

An important concept is the use of colors that are either similar or complementary (opposite) as a means of introducing harmony or emphasis, respectively. Color in graphic design is not our forte, so we will not attempt that instruction, but we will offer one piece of advice: try to stick to one color scheme in your presentation. It is easy on the viewer's eyes and gives an air of professionalism (in contrast to slide decks that are motley because they were assembled from slides from many different original decks).

TEXT AND FONTS

We can't leave the topic of slide design without discussing fonts. For an elegant business presentation, font choice is extremely important. But it is also useful to pay attention to your fonts when designing a science presentation. You will find many pages on the topic in Nancy's book. Learn the basics. Choose a font and stick with it—multiple fonts on one page look sloppy. It looks like you cut and pasted material onto the page—which you probably did—so clean it up. We recommend a clean sans serif font (such as Arial, Calibri, Avenir, or Century Gothic, to name a few) following the argument of minimum non-data ink. But there is clearly no big advantage to using any particular font or font class. Use one you like that isn't hard to read. Be sure it is a standard font or you run the risk that it won't render correctly if your presentation is viewed on a computer that doesn't have the special font installed.

Of much more importance is the question of how much text to put on a slide. Garr Reynolds invented, and Nancy Duarte has popularized, the concept of a "slideument." This is a slide deck *that is intended to be read rather than seen in a presentation.* Clearly, the rules for such a deck are different than if you were presenting the material to a live audience. Often, however, we fall prey to the desire to make a presentation meet both standards: a slide deck that will deliver great visual impact but be completely understandable to someone reading it later. This is a difficult task.

A viewed presentation has different requirements for text than one you read on your own. In the live version, text serves to give the audience a sense of what the topic is or to highlight a key finding that you want to stick in their brains (very much like the "renewables" arrow in our California electricity graph in figure 11). If you have a set of points that you want to lay out for the audience, simple lines of text are a good way to do it. But avoid the temptation to fill in additional elements below each bullet—no one has time to read that. If they do read it, they won't be listening to you. Despite all of our best efforts, it is just not possible to listen and read at the same time. Even switching back and forth quickly causes us to lose content. Don't ask listeners to do it—keep your text simple.

The slide images in figure 14 imitate what your deck looks like from the back of the room. It's not possible to read the sub-bullets in the slide

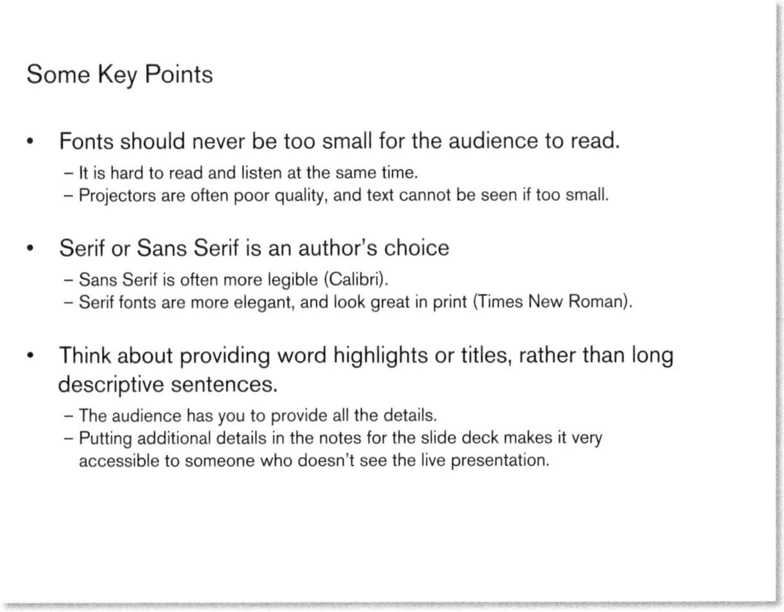

Some Key Points

- Fonts should never be too small for the audience to read.
  - It is hard to read and listen at the same time.
  - Projectors are often poor quality, and text cannot be seen if too small.

- Serif or Sans Serif is an author's choice
  - Sans Serif is often more legible (Calibri).
  - Serif fonts are more elegant, and look great in print (Times New Roman).

- Think about providing word highlights or titles, rather than long descriptive sentences.
  - The audience has you to provide all the details.
  - Putting additional details in the notes for the slide deck makes it very accessible to someone who doesn't see the live presentation.

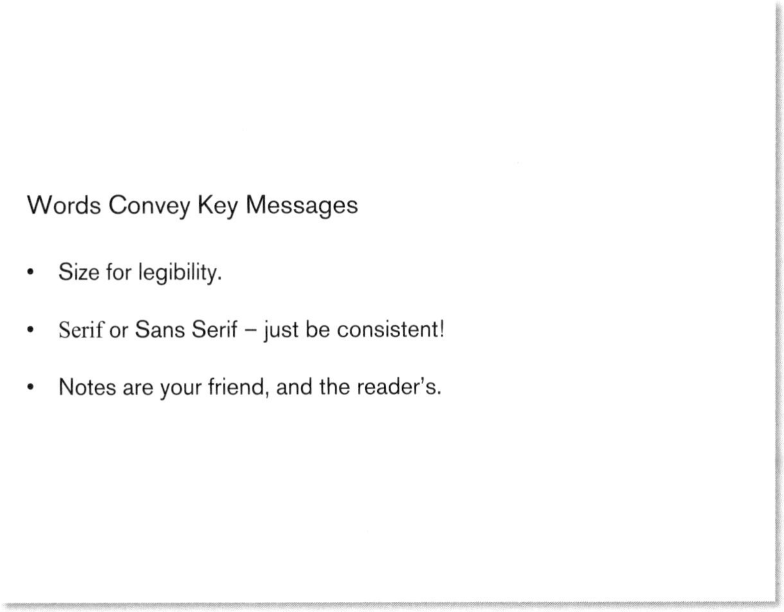

Words Convey Key Messages

- Size for legibility.

- Serif or Sans Serif – just be consistent!

- Notes are your friend, and the reader's.

FIGURE 14. *Top*, an overly wordy and illegible text slide. *Bottom*, an improved version conveying the same information.

on top without making a lot of effort—and they duplicate what you are probably saying anyway. The version on the bottom understands that most viewers will look only at the high-level bullet points and is respectful of the audience by keeping the points short. They are headlines for what you are saying.

The bottom slide takes advantage of another method that scientists should follow—the title conveys information rather than simply serving as a heading. In scientific publications, we use bland headings like "Discussion" because the reader expects there to be a discussion section and often wants to page ahead to find it. In a presentation, the reader doesn't have the option of paging ahead.

Instead of using "Discussion" as a heading, take the opportunity to give one snippet of what the discussion *is*. Particularly when your audience is having a little trouble following your argument, a headline that states what you want them to understand from the slide is like a life preserver. This is even more important when the slide deck is to be read later—it is often very hard to extract the key takeaway messages when reading a slide deck. This is one of the presentation principles that applies no matter how you will present your material. Make the title matter. Let it help you tell the story.

The important concepts are the same whether you are using bullet points or titles or highlighting things on graphs and visuals. Use large, legible text, and think *headline,* not *detail.*

"But," you say, "The key decision maker couldn't make it to my talk! She is going to have to read the presentation later." At this point, you have a tough call to make. You could capitulate, create a slideument that is only meant to be read, and give a really terrible seminar to all of the key decision maker's staff. Wow, that sounds like a great idea; they will be so respectful of you ignoring them in preference to the boss, who isn't there! But there is a better answer—pass out your talk in the form of notes pages. You can distribute them in PDF format, and they are delightful to read as standalones. Nancy Duarte is a strong proponent of this approach, and as the author of entire books that were created as slideuments, she is certainly an expert on the topic.

Figure 15 shows what a notes page looks like—there's plenty of room for text that can remind you what to say and fill in the details for anyone reading the handout later.

If you feel that a slideument is really called for, another way to avoid making two separate presentations is to insert text slides in between the formal presentation slides. You can turn them on or off depending on

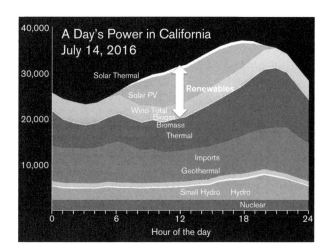

24 hours of power in California on the day I wrote this talk – nothing special about the day.

Imports are from the Western power grid, which contains some coal, although California no longer enters into new long term contracts for coal power. Imported power must have lower carbon footprint than a good natural gas plant.

Solar power is only utility scale systems – does not include rooftop, which is 'behind the meter' and doesn't show up as a source, but instead a reduction in needed electricity.

Units are megawatts

Data source http://www.caiso.com/green/renewableswatch.html

FIGURE 15. A notes page is an excellent way to distribute your slides with descriptions that avoids putting all the words into the presentation version of the slide.

the situation. Just don't give in to the temptation to leave them in when you are presenting!

KEEPING YOUR PRESENTATION FRESH AND UNDERSTANDABLE

Asking your audience to spend time looking at your slides is a social contract. In exchange for their attention and potential support, you agree to respect their time. Apart from our advice on the scientific part of that contract, you also should pay attention to simply keeping your listeners engaged and alert.

Any of the good presentation books, including those by Reynolds and Duarte, will emphasize that people learn in chunks of time no longer than ten or fifteen minutes. Try to keep your presentations organized into chunks no bigger than that, with clear demarcations in between the sections. For instance, if you are using typical white-background slides, all-black slides with simple titles can be used to set off the sections of your talk. This tells the audience that you are moving to a new topic and that their brains can relax a little on the previous topic.

Susan Weinschenk, the psychologist we introduced in chapter 6, "The First Two Minutes," is the author of a useful book series, *100 Things Every Designer Needs to Know about People,* where she lists one hundred tidbits from psychology and motivation that apply to design or presentations. Susan provides an outstanding list of recommendations and reasons why. While the book is aimed more at business, she gives us enough background to see how her comments apply to science. People need context; people have mental models; people process information best in story form; people learn from examples. Sound familiar? All of these apply to any context.

Keep It Bite-Sized

The very first thing in Susan's design bible is: keep it bite-sized. The brain can process only a small amount of information at a time—consciously, that is. (The estimate is that you handle forty billion pieces of information every second but only forty of those make it to your conscious brain.) One mistake that presenters make is giving too much information all at once.[10]

Of course, this turns out to be the basis of the length of TED talks. They are limited to twenty minutes because that is as long as people can pay rapt attention to anything. If you want to be sure you are going to have impact, limit yourself to twenty-minute chunks. Editing is the essence of impact. *Leaving things out highlights the things that stay in.*

The twenty-minute rule is a tough one for scientists. There are certainly championship opportunities that are longer than the typical decision maker discussion we have focused on. What about a formal seminar? The twenty-minute rule still applies, but in this case you should break your talk up into segments. If you have a demonstration or a gadget to pass around, do it at the twenty-minute mark. If you don't have a physical break, give people a signpost with a really clear slide that denotes a new section of the talk. An excellent way to break up a talk is with a visual interest picture of your colleagues working—it gives you a chance to talk about something that is topical but very different than your main subject. And it is a terrific way to acknowledge collaborators, much better than just saying their names.

Address Experts and Nonexperts

When you are speaking to a mixed audience of experts and nonexperts, it is often a great idea to address both levels of knowledge. Josh White, whose quotation opens this chapter, is a winner of the prestigious Lawrence Postdoctoral Fellowship at Lawrence Livermore National Laboratory and a true artist when it comes to working with groups that include decision makers and his scientist peers—a tough mix when you are trying to determine the appropriate level of detail in your discussion. One of his signature techniques is to use a parallel approach. Figure 16 shows one of his slides. One side is in plain English, while the other shows a complicated equation. The plain English describes what the equation does, for the benefit of the decision makers, and the equation itself satisfies the scientists in the audience. The slide also shows how you can use a tiny header line running along the top to show the overall outline of the talk and highlight the current section to indicate progress. It can be very comforting for listeners who are in over their heads.

We can't finish this chapter without a brief discussion of what to leave *out* of your slides. Our corporate and academic masters are proud of the expensive graphics and logos they have designed and often expect us to include them on every slide. With all due respect to the branding team, to do so is to drown your audience in slide junk. Slide junk is one of the most insidious problems a scientist can face in preparing clean visuals. All the lines, template items, logos, colorful graphics, and organizational identifiers that everyone seems to be fond of, and that

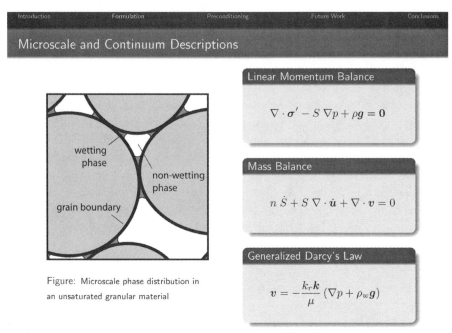

Figure: Microscale phase distribution in an unsaturated granular material

FIGURE 16. An example of a slide for a mixed audience, containing both nontechnical (the image, *left*) and technical (the equations, *right*) explanations. Figure by Josh White.

slide software manufacturers love to add as if they were enhancements, detract more than they add to the typical presentation. Why would you need a purple wave on every slide?

Too often you will see a logo from your organization, a logo from the sponsor, and the title and location of the meeting on every slide. This is completely unnecessary. These items can be respectfully placed on the title slide or reproduced on the final slide as a form of acknowledgement, but surely everyone knows which organization you are from after the first slide. Craig Hadden, an Australian presentation expert, has an excellent idea for your corporate logo: show it once on a completely black background at the beginning of your talk and say nothing when it is on the screen—logos are supposed to speak for themselves after all![11] This creates full impact without distracting from your material.

If you are worried about your slides being copied without attribution, use the footer space to give your organization's name. As long as you are using that space, add the date as well—you will thank yourself for this later when you wonder just which version it is. And always, no matter what, use a small page number in the footer space, but feel free to let any graphics or photos share that space. Referring to slides by page number is a valuable tool when you have to present information without everyone in the same room. If your organization throws a fit about slide formatting and insists that you include all their corporate elements on every slide, feel free to put your graphics over them. Oops!

PART THREE

Honing Your Communication, Influence, and Emotional Intelligence Skills

10

Improving Your Speaking Skills

Even the most polished communicators can get better. As you reflect on what you naturally do well, you'll find there are always areas where you can up your game—and your confidence. In this chapter, we look at several elements that form the foundation for your presentation skills as well as techniques for effectively handling Q&A sessions after a talk or during a poster session.

USING YOUR VOICE

Your voice does most of the heavy lifting when you communicate. It is an instrument, and you can optimize its sound and effectiveness by learning to "play" it. We can't all be lucky enough to have James Earl Jones's commanding voice, but skilled communicators pay close attention to how they sound, starting with volume.

Be loud enough to be heard. That goes without saying. But how do you manage that in a large room? Borrow techniques used by singers. They stand tall, breathe from the diaphragm, and open the back of their throats to let their voice resonate. You should also check in. Ask the audience, "Can those of you seated in the back of the room hear me?" If the answer is no, take deeper breaths to get more of a foundation for sound to help you increase the volume of your voice, or ask to turn up the volume on the microphone. Always use a lapel microphone if it is available. Wearing one will ensure your mouth is far enough away to avoid the distracting sound of a mic that is too close.

Vocal Variety

Simply stated, vocal variety refers to the differences in pitch, volume, and pace that help communicate meaning. You may have a fabulous idea or promising research, but you will fall short if you drone on in monotone when you tell people about your work. Varying your voice helps bring your message to life in a way that captures the listeners' attention.

Listen to your favorite songs and you will hear what vocal variety sounds like. Pay attention to the way the singer changes pitch, volume, pace, and phrasing to carry the message behind the lyrics. Radio personalities and actors use vocal variety well, too. Typically, they speed up to display enthusiasm. Likewise, they turn the volume all the way to loud or soft for emphasis. These changes keep people listening. Just like a great singer, you can use variety to underscore your message and grab attention when you want people to focus on what you say next.

How can you incorporate these techniques if they don't come naturally? With practice and design. Think through the points you want to make and where you want to draw extra attention. Find the words or phrases that should be loud or soft. Rehearse aloud. Record yourself so you can listen for places where you can speed up or slow down for emphasis. Find catchy ways to say things. Short sentences are powerful. Repetition is, too.

Pause for Emphasis and to Stop Saying "Uh" and "Um"

Pauses are another powerful speaking tool. A short pause may seem like a long time to you, but it gives listeners a chance to digest information and is much appreciated. A carefully placed pause shines a spotlight on the words that follow it. Pause between paragraphs. Pause between sentences. Pause when you want people to pay full attention to what you are about to say. The other place that pausing is important is when you are changing topics. Give the listener a quick mental break before you take their attention in a new direction.

If you are in a forum where you have to turn your back to your audience to change your slides, keep in mind that it is far better to pause and wait until you are facing your viewers before you start speaking again. That way your voice doesn't get muffled.

Pausing is also a great antidote for those of us who find we say "um" or "uh" too much or fill silence with other nonwords. Learning to pause to collect your thoughts will make the need for nonwords disappear. You can see that pausing has many benefits. Give it a try!

GESTURES, FACIAL EXPRESSIONS, AND MOVEMENT

Effective use of gestures can catch the eye of your listeners and get them to focus on you. Gestures convey excitement and passion. When your gestures and facial expressions reinforce your conviction, your listeners get engaged.

Gestures are remarkably authentic—if you can let yourself relax. Most of us will find that our hands and arms appropriately illustrate concepts if we don't force movement or constrain our bodies. You just have to trust yourself and pretend you are simply talking to a group of people you know—even if you are standing before a sea of five hundred unfamiliar faces.

If you want to see what it looks like when speech, facial expressions, and gestures align, watch the presentation of Three Minute Thesis winner Dustin Chernick from the University of Minnesota on YouTube.[1] We met Dustin in chapter 7, "Crafting Key Messages and Narratives,", where we discussed his use of iconic analogies. In his talk, titled "Using Good Cholesterol to Treat Alzheimer's Disease," you see him appear relaxed and poised. Early on he makes an expansive, sweeping gesture with his hand as he moves from explaining that the number of people with Alzheimer's will rise to over fifteen million by 2020 to emphatically proclaiming, "We don't have a single drug that can stop it."

Dustin appears passionate and genuine, and his gestures reinforce the meaning of his words. Halfway through his talk, he explains, "My research is focused on the connection between the heart and the brain, and there is a big one." He moves his hand from his chest to his head and points at the audience to draw our attention to the word *big*. He also emphasizes the word *big* with his voice a bit. This important alignment between what a listener hears and sees adds to the speaker's credibility.

Keep in mind that in large venues, it's important to exaggerate your movements and make your gestures big enough to be seen in the back of the room. Walk around and get closer to your audience. Don't stand behind the podium. If you are on a stage, simply moving in the audience's direction is enough to get them focused on what you are saying.

Facial Expressions

The look on your face is a very important part of nonverbal communication. Your smile and your eyes are two powerful elements that help show emotion. When your facial expressions are in sync with what you are saying, you are amplifying your communication and making your words more credible. To enhance your ability to use facial expressions to convey an idea or feeling, spend time in front of a mirror so you can see what you look like as you speak. Consider where it will be important for your facial expressions to reinforce your message. Be aware of the phrases where you might want to open your eyes wide or break into a smile. Take advantage of smartphone technology to see yourself in action and self correct.

Movement

Watching someone pace back and forth across the stage or shift weight from one foot to the other distracts from what is being said. Instead, stand comfortably relaxed and use good posture. Don't strike a pose that stands out. Clasping your hands together as if in prayer or crossing your arms across your chest—in what's typically seen as a closed position—are two common stances to avoid. It's best not to constrain your arms in any way. Instead, let them hang naturally by your side so they are readily available for gesturing.

Moving to different areas of the room or stage can be a very effective way to engage your listeners, as long as you move with purpose. Change your location at the end of a segment of your talk or when you are switching to a new topic. Walk during the time it takes to say a complete sentence. Then stop. Stay put for a few minutes. Deliver the next few paragraphs or the entire section. Then move again. Staying locked in place is far less engaging. Moving at appropriate times works. Moving toward the audience can be especially effective when a question is asked.

LASER POINTERS

The laser pointer is not a cat toy. But your audience's eyes will follow it just like a cat's do. Humans evolved from hunters, and our eyes cannot resist tracking movement in a quiet room. It is always best to keep your slides self-explanatory, but when detail requires a pointer, use it sparingly. Hold it with both hands, and don't turn it on until it is pointed correctly. Don't circle the pointer in the vicinity of the item you are highlighting. Hold it steady. This is a prime opportunity to keep your audience focused on your message. Roger would tell you that when the pointer strays, you are setting yourself up for any one of these likely audience reactions: "Am I supposed to see something there?" "Ooh, shiny!" And of course the ever to be avoided, "What an idiot."

An old-fashioned wooden pointer has merit. It keeps you correctly oriented, close to the screen, but not blocking it. A stick is unlikely to wander as much due to its weight and our long experience poking things—it is part of our muscle memory. The challenge with a wooden pointer is finding a secure place to set it down after you've put it to use.

HANDLING QUESTION-AND-ANSWER SESSIONS

Most of the time, a talk isn't over until the speaker takes questions from the audience. Typically, ten to fifteen minutes are allotted for a Q&A session. You will want to make time for questions in small group meetings, too. How many questions should you be able to answer in that time? As many as possible.

What one person is curious about may not matter to another. To keep everyone interested, you need to understand how to make the most of the dialogue and not fall into the trap of engaging far too long on any single question.

Science champions view the Q&A session as a valuable opportunity to reinforce key messages, make a connection, and ultimately drive action. Learning to respond to questions effectively is a powerful skill that will serve you well whenever you encounter a curious mind.

You don't want to be the speaker who starts in on an answer and burns five minutes providing too much detail. Unfortunately, however, that's a common occurrence because scientists love to discuss specifics. Don't go there. Learn how to elevate the level of your response so it is more effective. Think of your response as a chance to give the headline and first sentence or two of explanation. Beyond that, you are likely to go too far.

Listen Carefully

Keeping your responses succinct requires you to listen carefully so that you hear the intent behind a question. You may need to filter through the preamble before you hear the actual question. Identifying the heart of a question helps you know how to approach the answer. Resist starting to formulate the answer in your mind while the questioner is still meandering on the topic. Stay focused and listen to what is really being asked.

Reframe the Question before You Answer

Let's say you get a question that has to do with the latest findings on using immunotherapy to fight certain cancers. You may be very familiar with a number of studies and could provide a very detailed response. Instead of diving in, begin your response by reframing the question. You could say, "You asked about the latest findings on the effectiveness of immuno-therapies in treating cancer." Now you have a broad question that is focused on effectiveness. Your response can be a high-level characterization of overall effectiveness. You can give a specific example if you know the relevant data, but you should not attempt to compare and contrast multiple studies because it would take too long. You might reference Jimmy Carter, who was treated and cured for a metastatic melanoma with Merck's Keytruda drug, as a concrete example of the success that's possible.

Sometimes you will be barraged with a complex question that has multiple elements. Taking a moment to reframe the question enables you to collect your thoughts before you respond. It also makes it possible for you to hone in on the aspect of the question that you want to address in your answer. Reframing the question and restating it before you start your answer is appreciated by the audience. It is vital if you are in a situation where the person asking wasn't loud enough to be heard.

Once, after giving a dinner talk about climate change and fracking, Roger was asked about the elements in the wastewater, the risk of contaminating drinking water, and whether injecting water back into the ground causes earthquakes. He set up his response by saying, "You've asked about several of the challenges related to the water used for fracking. I only have time to address one, but I'm happy to connect after the talk to respond to all of your questions." This reframing step helps reassure your listeners that you are not about to take up all the Q&A time responding to one long-winded questioner. And, it gives you the opportunity to invite action and welcome a connection with members of your audience. If you are in a small group discussion or talking to an individual, you can handle a multipart question like this by acknowledging the parts of the question and asking the listener which topic they would like you to address first.

Keep Everyone Engaged

As you listen to the initial question, it's essential to make eye contact with the person asking it so you can keep your attention focused on their inquiry. Once you begin answering, be sure to shift your gaze to others in the room. By doing so, you open the conversation to the entire audience. In a larger forum, the goal is to avoid the sense that you're having a conversation only with the person who asked the question, who might then feel comfortable asking a follow-up question. In a smaller venue, it is more acceptable for one or two individuals to dominate the questioning. Be aware of how everyone in the room is responding and try to avoid situations where disinterested individuals get stuck in conversations they don't want to be in. Offering to continue the conversation with the questioners and giving the others an opportunity to leave the room and is always appreciated. You might say, "I can tell you have a number of questions, and I am happy to stay and answer them for anyone who would like to join us." Say thank you to the people who opt to leave and then go back to the conversation with your rapt listeners.

Some questions are best answered offline. When you hear a reference to a specific situation, institution, or study, it's likely that the conversation will end up needing a one-on-one response. Consider this an opportunity to engage at a deeper level and take a two-pronged approach. Give a high-level answer and then offer to go into more depth in a follow-up conversation. This makes it clear that you are knowledgeable about the subject and shows respect for the audience at large. Invite the questioner to provide their business card after your talk so you can contact them directly. Leverage this opportunity to engage with people who have a sincere interest in your topic and demonstrate the potential, through their question, to take the action you desire. While you may not know who is doing the asking, it's safe to assume that apparent technical interest warrants follow-up.

Reinforce Your Message

Most questions serve as a jumping-off point for reinforcing your messages. There are many ways you can link your answer back to your topic. Simple phrases like, "And remember," "That brings me back to the point of my talk," "This is another reason why," and "Fundamentally we find that," are known as bridging phrases. They enable you to deftly weave your way back to the point you want to drive home.

When is it appropriate to go into great detail during a Q&A session? If you are presenting at a forum with an audience of technical experts who share a common understanding of your topic, you can get into more technical depth. You want to check in with your listeners to be sure you are responding at a level that meets the needs of their inquiry. You might give a response and then say: "Does that give you the information you need? Is there anything else you want me to elaborate on?" You can use this technique a few times to be sure you satisfy your audience's appetite for information. It works well with decision makers, as it helps ensure you fill any information gaps that might be keeping them from taking action.

Invite Questions

How do you get the ball rolling if no one wants to be the first to ask a question? Start with being very aware of your body language. Don't stand back with your arms crossed against your chest. Instead, stay open, smile, and say with enthusiasm, "I'm happy to answer any questions that you have for me now." Wait for at least three seconds. If you don't get any takers, you have two options: reinforce your key message, thank your listeners, and leave the stage; or give people a little more time to formulate a question. Do so by posing and answering a question that you think may be relevant: "I'm often asked . . ." This approach can be all that the audience needs to get engaged. It also gives you a ready-made invitation to reinforce your message.

If your attempt to get the questions flowing doesn't do the trick, it's time to exit the stage. Don't just deliver a perfunctory thank you. Close with emphasis on the message you want to have everyone in the room take away. "The TOPIC holds the promise of solving the PROBLEM. We are excited about the findings in our early studies and welcome the opportunity to further our research or work with organizations to find new applications of our SOLUTION. I appreciate the opportunity to tell you about our project today and invite you to contact me or provide your card so we can continue the conversation." This kind of call-to-action close is the ideal way to leave your topic when you are presenting to a broad audience. In a small group session with important decision

Handling Questions during a Poster Session

Poster discussions require you to converse at a more detailed level. That said, you still want to parse your response so that you only go as deep as needed. Phrases like, "Does that answer your question?" help ensure you've satisfied your listeners.

What's the best way to handle those awkward social moments when you are deep in conversation and someone else approaches? Acknowledge the newcomer by saying, "Jennifer was just asking me about X, and I was telling her . . . " When you finish answering Jennifer's question, turn to the newcomer and ask, "Is there a question I can answer for you?" This approach gives Jennifer the choice of sticking around or moving on. If you'd rather not interrupt your conversation with Jennifer, signal that to the newcomer by saying, "Jennifer and I will need another five minutes to adequately address her question. You are welcome to listen in, or you may want to come back a little later." This is a courteous way of taking charge of the situation and setting clear expectations for everyone involved.

makers, you'll likely want to make a more definitive ask and seek a response before you conclude the conversation.

Prepare for the Q&A

If you are giving your first talk on a topic, you would be wise to do some strategizing about the Q&A session. Think about your talk as a pathway to questions. How can you construct your talk to elicit the kinds of questions you will be prepared to answer? Brainstorm a set of possible questions and be sure to practice your answers so you know what you want to convey if asked. If you have time, do a dry run for a friendly and informed audience and ask them to come up with questions. This will help you anticipate what listeners might be curious about.

As you think about the best way to construct your answers, look for opportunities to reinforce your key messages. This is premeditated bridging. Look for analogies or examples that will help simplify complex concepts or bring your message to life. Don't expect to come up with these on the fly. By doing your homework and contemplating likely questions, you can make the most of the Q&A session.

Escape Common Q&A Pitfalls

Avoid saying, "That's a great question." Yes, it's a filler phrase to give yourself think time, but if you use it, you subject others to feeling inadequate when their question doesn't get the same response. It's okay to simply thank a person for asking or say, "I get that question often," or, "I can appreciate why you are asking," but refrain from complimenting some questions and not others.

What if you are not sure of the answer? It's better to say, "I don't know," than to guess and be wrong. You can commit to getting in touch later with a researched answer or to posting one on your website, but don't feel pressured into answering if you are not sure of the facts. If you can answer a related question, offer to do so. You might say, "I don't know the answer to your question, but I do know this . . ."

What do you do when you get a question that escalates into a heated exchange? It's easy to get sucked in. But it's best to remain diplomatic at all times. Show interest in continuing the dialogue, but be clear that you will do so offline. You could say, "Clearly there is a lot of energy around this topic and some strong and differing opinions. I would like to continue to discuss this after I conclude my talk, so let's plan on that."

Debating in a large public forum rarely results in a desirable outcome. If you wait to take a topic offline, the heat usually dies down. If it doesn't, this is your opportunity to really listen to what's behind all the emotion and see if you can close a paradigm gap. Be open to learning why someone has a radically different point of view. In the end, the mature response may be just to agree to disagree once both sides have been heard. If this is happening with a key decision maker, you'll have work to do to understand their concerns and see if you can adjust your approach to win them over.

11

Influence and Persuasion

Influence. An action exerted, imperceptibly or by indirect means, by one person or thing on another so as to cause changes in conduct, development, conditions, etc.

Persuasion. Successfully urging *to do;* talk *into or out of* an action; attract, lure, or entice to something or in a particular direction.

—*New Shorter Oxford English Dictionary*

Championing science may feel like an act of persuasion—and ultimately it may be. But often that persuasion is the culmination of a long series of acts of influence. The art is in gently pursuing the path of influence to pave the way for persuasion rather than attempting to persuade in a single interaction.

Creating influence and being persuasive with decision makers takes a lot more than citing facts. When your listeners walk away from their time with you, you will have succeeded if you have made three key emotional impacts. The audience should be *excited* about what you are doing, *trust* that what you have told them is true, and most importantly, *be able to act* on what they heard. You don't want to leave anyone frustrated because they love what you said but don't know what to do as a result. Let's take a closer look at these three important elements.

GENERATING EXCITEMENT

Science is not about the banality of the everyday. When you champion science, your passion is the first emotion to display. That's why we encouraged you to put passion into your first two minutes—it conveys your excitement about the topic. But your goal is more than just to have a good

---

The Importance of Why

Recognize that champions talk about why the science matters, while the conventional academic approach is to talk only about how the science works. Making this shift doesn't mean there is no longer a need to explain how your science works. It's a matter of how you structure the conversation and what you cover first. You can transfer your passion to the audience if you do a good job of explaining *why* you are excited.

---

start. You want your listeners to end up as enthusiastic as you. You want them to leave your talk and tell someone about what they heard. How do you generate that kind of excitement? Paint a picture of what is possible and how your science will advance your cause. Explain why your science will make the world a better place or make a difference to people. Make your audience feel that they can help make that future happen.

THE IMPORTANCE OF BUILDING TRUST

Since decision makers are rarely experts in your topic, your challenge is to present yourself in a manner that leaves them understanding and believing what you say. Earning the trust of a nonexpert audience is difficult because scientists are trained to talk to experts. And, as Randy Olson tells us, trust is vital.[1] He believes that the audience must trust you, or else no amount of information will be able to convince them of your thesis.

Trust runs very deep in our neural makeup. Our brains are configured to rapidly assess situations that represent danger or opportunity. This hardwired ability protected us from predators and enemies. Today, perhaps the most obvious example of this is the way we assess emotions in other people. A simple glance at another person's face tells us whether they are happy or sad or, most importantly, how they feel about us at the moment. Are they going to yell at us or hug us? Even the least self-aware scientists usually can tell in an instant—we have had enough experience with similar situations that our brains quickly identify the correct response.

How People Make Decisions

Rapidly processing facial expressions is an example of what Daniel Kahneman calls *fast thinking* in his excellent book about how people make decisions, *Thinking, Fast and Slow.*[2] Kahneman is a psychologist with the distinction of having won the Nobel Prize in Economics for his work on trust and decision making. He contrasts the thought process required when we are asked to solve a difficult problem, like mental math. In this case, the answer doesn't just jump into our head—we have to deliberately align our thoughts to perform the calculation, recalling from memory the methods to do it and probably excluding a lot of other thoughts from our head while we perform the task. This is *slow thinking,* and of course it is at the heart of much of what scientists do. Slow thinking is deliberate and often intense. Fast thinking is automatic, requiring no conscious control and taking very little attention.

Kahneman's book is important to us as science champions because he describes how many decisions are made, inappropriately or not, on the basis of fast thinking. When presenting an idea to a decision maker, we have to keep these two systems of thinking firmly in mind—because they can either help or kill our efforts.

> System 1 (**fast**) operates automatically and quickly, with little or no effort and no sense of voluntary control.
>
> System 2 (**slow**) allocates attention to the effortful mental activities that demand it, including complex computations. The operations of System 2 are often associated with the subjective experience of agency, choice, and concentration.[3]

Kahneman gave a very influential lecture concerning the direct application of these two systems in science communication at a 2012 symposium called The Science of Science Communication.[4] In the lecture, he summarized how to apply these two systems to the way we approach different audiences.[5] When your listeners are not experts, you are much better off speaking to their guts, to their hearts, to their System 1. And this is best done by describing concrete events and telling stories. In Kahneman's words, System 1 is more "bound to the *coherence* of the story than the evidence behind it." Moreover, because the story you tell needs to be emotionally coherent, as we saw in the discussion of information deficit theory in chapter 5, "Why Scientists Communicate Poorly outside Their Field," the source of the story is extremely important. You must be trusted. Kahneman says, "Messages from a distrusted source will be ignored, and the amount of evidence will not matter."[6]

Go back and read that last line again. It is at the heart of championing science.

How to Build Trust

In general, building trust is done by consistently demonstrating that you are reliable, accountable, and honest. Scientists can also build trust through association and peer networks (attended Caltech, worked with Professor X, etc.) and of course through publications, which are reviewed by a sort of jury trial to confirm the validity of the findings. Sharing valuable information also helps build trusted relationships, especially when the information is helpful to the decision maker. In person, making appropriate eye contact also helps people feel they can trust you. The most significant nonverbal driver of credibility is how well your tone of voice, gestures, and facial expressions match your words.

Another key dimension of building trust is effectively establishing your credibility. Credibility is built by delivering the work you promised in your proposal and by keeping your sponsors informed about progress through transparent communication.

Marine biologist and moviemaker Olson agrees with psychologist Kahneman: the audience must like you and trust you for your message to easily be accepted. The first and foremost rule in this regard is, don't begin by alienating your listeners with material they don't understand. Meet them where they are, and bring them along with respect for their knowledge and experience. But interestingly, it is also valuable to just once really dive into the deep water of your science.

The Importance of the Deep Dive in Building Trust

Almost every audience deserves one deep dive, one statement or slide that shows you really know your stuff and that would be recognized as such by any expert in your field. Why? First, there may be experts in your audience. Second, nonexpert listeners *expect* you to present material that is over their head. Just be careful not to drown them with it. Meet their expectation, and that of any experts who may be evaluating you, with a well-crafted statement for one of the important scientific issues in your material. Gauge the material based on the audience, as discussed in chapter 4, "Who's Listening?"

For instance, if you are pitching a model to a governmental agency, you might want to throw in the differential equations that control the response. Josh White's approach in figure 16 (see chapter 9, "Designing Effective Visuals") is a terrific example of this. It addresses both the math and the physical phenomena that he was modeling.

An effective example of a deep dive for a very nontechnical audience can be found in the Three Minute Thesis presentation by Kiara Bruggeman, an engineering and computer science student at Australian National University.[7] In this case, the definition of *deep* is of course much different than for Josh's audience. In her talk about rebuilding brain tissue after a stroke, she defines the extracellular matrix, which is a key concept that underlies her work. Kiara provides an easy-to-follow explanation, which demonstrates the depth of her knowledge yet uses words that are accessible to her audience: "In healthy tissue, cells exist in a fluid environment called the extracellular matrix. This matrix is full of useful proteins and also offers structural support in the form of nano-fibrous proteins. Different tissues have different needs, so every matrix is unique. I've been making materials that mimic the nano-fibers and matrix specific to healthy brain tissue."

She clearly demonstrates that she has the requisite expertise. She also builds trust by packaging the information in a way that conveys expertise but does not overwhelm her listeners with concepts they can't comprehend.

The deep dive might seem like a difficult concept to get exactly right. But as long as you restrict yourself to just one deep dive in any circumstance, it doesn't matter that you might lose some of your audience. Just make sure that the next topic you discuss is described in ways they can comfortably understand.

Gaining Funders' Confidence That You Will Deliver

When you ask a sponsor or investor for money, they always want to be certain that you are going to deliver the product you promised, whether it is new science or a new business. Building trust requires you to demonstrate that you have the right team and you know how to manage the undertaking. Nancy Floyd helps us appreciate how essential this criteria is for venture capitalists: "I think most every VC would agree they would rather have a B technology and an A management team versus the other way around. It comes back to management, management, management."[8]

She cautions funding seekers to appreciate the level of commitment they need to demonstrate to build trust:

> Specific to deals coming out of labs and universities, we want to see that there is no safety net, meaning the team doesn't come in and say, "Oh, we are taking a five-month sabbatical from the lab and we're going to try this but if it doesn't work, we get our jobs back." We want to see that the team has put everything on the line, including their jobs, benefits, etc. In other words, that the team is 100% bought in. So, in a real sense you've got to cut your economic ties, though not all of your other ties. We've had plenty of companies continue to use resources at the labs and universities. In fact, we encourage that for a while. It keeps the costs of starting the business low, upfront. But, we do want to see that the management team is equity driven and will do absolutely everything they can to make their company successful.

Finding the Right Approach for Disclosing Concerns

Nancy's advice raises an important question about trust for scientists to consider carefully. When you are working to build trust, there will be things you think are true but aren't certain of. How do you strike the balance between disclosing everything that you are worrying about—as in telling someone everything that might be a negative—with being optimistic about the future? We asked Nancy how she prefers scientists or technologists to balance conveying what could fail with the promise of success, and we think her response applies no matter whom you are approaching for support: "I like to know that people are realistic. But frankly they also have to be optimists. Scientists have to believe that even if there are significant technical hurdles ahead, they're going to be able to knock them down. They've got to believe they are going to succeed. I like to hear people say, 'We know these are hurdles and have thought about ways to deal with them.' Confidence yes. Overconfidence no."

Nancy acknowledges that there is "a lot you are just not going to know when you go into an investment as an early stage investor. You don't know whether the technology can actually scale from the lab to the field, because the entrepreneur who is more expert than anybody doesn't know that either. You can look at a technology roadmap and decide whether it's reasonable or not. But, there's no guarantee you will be right."

Funders are essentially making a bet on you, so the more you can do to earn their trust, the more likely it is you will receive funding. That means you must never inflate your credentials or claim accomplishments that won't be confirmed by a thorough background check.

ENABLING LISTENERS TO ACT

The champion talk is a call to action. Whether you are informing your homeowners association about an environmental issue or trying to get elected officials to create a major new program, there is something you want your listeners to *do*. And they would not be listening to you if they were not interested. If the decision maker is excited about what you have to say and trusts you, then they will be more inclined to act. The most valuable thing you can give them is a set of actions that work in the context of their world. Never *demand* action from an audience, but never miss the opportunity to *let* them act (influence versus persuasion). Make a request or a suggestion, or pose a question.

A call to action can be as simple as asking listeners to call their political representative, review a paper and post a comment, or tell their colleagues about your work. If you are asking for funding, be clear about how much you need and how it will be used. With new areas of science, it is often important to confine your ask to a reasonable and achievable project that will help you learn enough to prove the value of continuing to pursue your work. Be sure you are making the request of the people who are in a position to actually give you the funding.

Think about what the audience needs from you to be empowered to go out and take action. It is not just about the science—it is about knowing how to act on it. That may mean enlisting other supporters on your behalf, introducing the decision maker to your colleagues, or taking other well-planned steps to pave the way for the support you are seeking. Most investors and sponsors have restrictive lanes in which they fund—you may be more familiar with the constraints of your science agency, which can typically fund only the acquisition of knowledge, not business development. Take the time to learn about the extent of your audience's constraints so that you can make an ask that permits them to act within those parameters.

What does an investor need to see to be able to act? What prompts them to make an initial investment? Nancy Floyd tells us it's not black and white:

> For us, generally there has to be some proof of concept, like a prototype, because in five to seven years we need to see significant commercial traction. A prototype helps us evaluate how much

money we think it is going to take and how much time. If it is a really credible group, we might come back and say we know you asked for a million dollars but let's put in $250K right now and we won't price it, we'll make it a bridge loan to see if we can get past the first technical hurdle. If that's the case, then we're prepared to write another check for $1.75 million. If it's a good team, we'll listen and we'll be creative.

She also reminds us that venture capital dollars are largely for product development, not technology development: "Venture funding is for sales and marketing and building out that infrastructure. So there are a lot of great businesses that should never raise venture capital. They don't have to grow 200% a year for five years to be successful. I always tell entrepreneurs that if you get rejected by a VC firm, that doesn't mean you don't have a good idea."

SEEDING INFLUENCE THROUGH THE GIFT OF INFORMATION

Time is an enormous gift to you from a decision maker. In exchange for the time of someone you wish to influence, you must return something of equivalent value. What constitutes a scientific gift? It is *always* information. Scientists have the expertise and analytical means to create knowledge that the world needs.

The gift-for-gift aspect of influence and persuasion is deeply rooted in human culture. Marcel Mauss's classic anthropology text *The Gift: The Form and Reason for Exchange in Archaic Societies* helps us understand this link.[9] Anthropology has a surprising amount to say to modern science. Mauss argued that before there were markets (money, trade, and places to exchange goods), the most substantive structure in society was the giving of gifts, and the requirement to return a gift in kind. Webs of relationships and obligations were built up as the objects were transferred, and societal links grew larger as people exchanged gifts. He contended that originally there was no such thing as a gift as we might think of it today, as something for which no recompense is expected.

Our societal norms still encourage this kind of gift-for-gift behavior in social situations. Creating a web of trust and obligation is an intrinsic part of influence and championing ideas. In the initial stages of a relationship with a sponsor or decision maker, there is no money changing hands. It is important to build the relationship, starting with small acts and moving to larger ones, growing to the point of having real impact on major issues. This growth occurs through building trust, and an important part of that is the exchange of gifts.

In most preindustrial gift giving, it was appropriate to reciprocate with a gift of equal value. That is still true today, as we can see from simply trying to determine the right bottle of wine to bring to dinner. As scientists, when we interact with a new person or organization, our *knowledge* provides the opportunity to create the cycle of gift giving and the relationship web that Mauss identified. By providing a free (unrequested) object of knowledge or evaluation, you have established that you understand the nonmonetary economy at the core of our social relationships. By making those gifts meaningful, you create the structure of a substantive relationship.

Ideally, you can give information that the other party needs but does not have. The more factual it is, the better. Roger has been in many situations with policymakers where multiple scientists are discussing a pressing current topic. Some will begin the relationship by telling the policymakers what the scientist thinks the policy should be. Others, much more effectively, will present facts about the basis for forming policy before giving their opinion. They position the facts as the basis for a later conclusion. For instance, what are the observed disease rates in a clustered outbreak, and what sources are known to encourage such a cluster? You can probably guess which scientist gets more attention—the one that provides the gift. The other approach, which starts with outlining the policy that the scientist believes should be created, is the opposite of a gift. It comes across as a demand of the policymaker.

There is a cautionary note, however. Excessive gifts may be refused or accepted with discomfort because of the burden they create. The challenge is to nurture a relationship by creating a friendly obligation to reciprocate in the mind of the receiver so that they are delighted to do something for you because you did something for them. Particularly because a gift is made in the open (as opposed to a paid exchange of services), the receiver may choose not to accept it and the associated encumbrance. Don't overreach on the gift of knowledge, but be prepared to give gifts of information frequently.

FROM INFLUENCE TO PERSUASION: ETHOS, PATHOS, LOGOS

As we move from the art of influence to the art of persuasion, we see some important parallels with our earlier discussions and deepen our understanding of how to champion science.

Aristotle is credited with identifying the concepts of *ethos, pathos,* and *logos* as the foundation for persuasion in his book *The Art of Rhetoric.* Ethos revolves around demonstrating credibility, which is an essential step in building trust. Pathos is about eliciting an emotional response, which ties back to the need to create excitement. Logos is the use of reason and logic to make a convincing case. Let's examine all three.

Ethos

When you are perceived as knowledgeable and experienced, you are more persuasive. Tell your story in a way that lets your listeners get a sense of who you are and why you can be entrusted with the subject matter. They need to see your expertise. They need to glimpse your character. They need to get a sense of who you are as a person so they can connect to you at a human level.

If you are being introduced before you speak or are introducing yourself, you have the ideal opportunity to start building your credibility. Write an effective bio. Craft it with one goal in mind: to demonstrate that you know your stuff. Mention your years in the field, awards, major scientific credits, and anything that shows how you are generous with your work and expertise. You may also want to include your educational credentials, and number of publications and patents. These are key elements of a trust-building introduction. If you can get someone to use the bio when they introduce you, that will serve as a tacit third-party endorsement of your expertise and lay the foundation for establishing your credibility. Roger's speaker bio is included in chapter 15, "Exercises for Applying the Championing Science Skills," as an example to help you craft one of your own.

Then it's up to you. When you enter the room or take the stage, smile. Stand tall. Doing these two simple things enhances your credibility. Remember, you have only one chance to make a first impression. Your goal is to come across as self-assured without appearing cocky.

The rest of your talk will either add to your credibility bank account or drain it. The most credible speakers are people who can translate technical concepts into simple, understandable thoughts. Sharing past successes as part of your story and giving credit to your collaborators will leave your audience feeling confident about you. Humility is magnetic. Grabbing all the credit repels.

What you say is only one dimension of building credibility. Paralinguistics explores how information is communicated beyond words. Listeners form opinions of you based on your tone and body language. The most significant nonverbal driver of credibility is how well your tone of voice, gestures, and facial expressions match your words. Inconsistency erodes trust. If you say you are excited about your subject but your voice is slow and monotone and your facial expression is flat, it won't feel genuine. Keeping a smile and a lilt in your voice and picking up the pace of your speech conveys contagious excitement. Record yourself practicing or making a presentation to see how you come across, and self-correct accordingly. You'll see how an upright posture, a natural smile, and an engaged expression exude confidence.

Watch the opening of the Three Minute Thesis presentation given by social scientist Michelle Brown from the Institute of Child Development at the University of Minnesota.[10] Michelle, who won first prize at the university's 2016 competition, shows us how powerful communication is when your voice, gestures, and facial expressions match your words. She does a compelling job of setting up her research area by painting a succinct and attention-grabbing portrait of the impact of the problem. She uses the iconic image of dominos falling. Her gestures are natural and appropriate. Here are her introductory words:

> For too many of the four million babies born in the U.S. every year, what was supposed to be a childhood filled with joyful bliss is shattered by the experience of child abuse or neglect. This all too common violation of trust is responsible for an annual economic burden in excess of $100 billion in the U.S. alone. When children are maltreated, it propels forth this domino effect, knocking down healthy functioning in nearly every aspect of life. It puts them at risk for peer rejection, school failure, depression, juvenile delinquency, unstable romantic relationships, socioeconomic disadvantage in adulthood, chronic stress, and even cancer. All of this due to a child being maltreated, once! Now imagine if that child gets victimized again by that same caregiver, a cyberbully, or an abusive partner. Sadly, this is often exactly what happens. My research shows that 61 percent of people victimized as children get victimized again as adults. But there is hope, because we know that a child's past does not have to define their destiny.

Her noteworthy gestures serve as a fabulous example, especially later in her talk. At one point, she makes a sweeping X by crossing her hands and rapidly pulling them apart as she says, "But enough about the things that we cannot change." Later, she holds her hand up and moves it toward the audience as she says, "One person has the power *to stop it.*"

Amy has always said that the best speakers are the ones who can truly be themselves in front of a room. Author Amy Cuddy, in her book *Presence: Bringing Your Boldest Self to Your Biggest Challenges,* gives an explanation of presence that reinforces why this is the case:

> Presence emerges when we feel personally powerful, which allows us to be acutely attuned to our most sincere selves. In this psychological state, we are able to maintain presence even in the very stressful situations that typically make us feel distracted and powerless. When we feel present, our speech, facial expressions, postures and movements align. They synchronize and focus. And that internal convergence, that harmony is palpable and resonant—because it's real. It's what makes us compelling. We are no longer fighting ourselves, we are being ourselves. Our search for presence isn't about finding charisma or extraversion or carefully managing the impression we are making on other people. It's about the honest, powerful connection that we create internally, with ourselves.[11]

Decision makers who rely on their gut sense (Kahneman's System 1) will be more favorably inclined toward your position if you come across as genuine, caring, and committed to advancing your science to deliver its benefits to the world. Establish your credibility by weaving contextual phrases like these into your conversations:

> Over the past three decades, and during my fieldwork over the past four years, I've seen this pattern dozens of times.
> When I started my research in 1982, I was captivated by the idea that . . . was possible.

The other basis for building credibility is by association and showing people that you have a common interest and common attitudes about what is important. This is a reflection of your character. Your collaborators are an incredibly important part of this process. Decision makers are often dubious about people who go it alone. In our highly educated and connected modern world, it is extremely rare for anyone to have an idea that no one else has had. And it will sound like hubris to most decision makers if you state that you are the only one who has thought of your concept or who is capable of resolving the issue.

On the other hand, having a few collaborators builds your credibility. Collaborators who are outside your domain build the breadth of your argument: the classic mix is university, research lab, and industry. This combination will help make the decision maker feel that their favorite kind of science is represented and that you are being informed (and constrained) by appropriate input from outside your immediate area of experience. Obviously, the best collaborator is one who is known, and trusted, by the listener. We recount a great example in chapter 16, "High-Impact Examples of Championing for a Cause"—the MIT/Caltech collaboration on the Laser Interferometer Gravity-Wave Observatory (LIGO). Both institutions struggled to make this breakthrough on their own, but when they joined together, the U.S. National Science Foundation finally felt comfortable to make enormous investments in the project without fear that some major mistake would be made.

A case of special interest is when a field has a small number, say five, of known strong institutions or groups (including your own). If your collaboration includes two of these five, the decision maker will be comforted that you represent a diverse and significant amount of the expertise in the field. If you are able to arrive with three of the five as collaborators, the decision maker will feel completely comfortable to make a decision based on the merits of your proposal because they know with certainty that no other group can exceed your collective reputation in this field.

Interestingly however, arriving with four or five of the institutions can appear self-serving and monopolistic. When the decision maker has asked for your input on an issue, this is an outstanding position, but if you are making the proposal on your own initiative, it takes away the decision maker's ability to get a third-party evaluation. And that takes away an important part of their ability to act.

Pathos

Pathos encompasses the things you do to evoke an emotional response in your listeners that will make them more receptive to your message. Appealing to emotions is not manipulative. It is a basic premise of persuasive communication, and it helps facilitate a shared understanding of the issue and what is at stake.

Sharing stories about your experience with your science or the early uses of your science create this kind of appeal. Rather than simply presenting data as numbers, wrapping them into a story about a person breathes life into your presentation and makes it more memorable and compelling, like the example that started with the death of the firefighters at Yarnell Hill from chapter 7, "Crafting Key Messages and Narratives."

Science champions can benefit from making use of five emotional appeals identified by Aristotle: anger, fear, shame, pity, and kindness.

Here's an example of how these emotions can be stirred when seeking support for a new approach to treat Alzheimer's:

Anger can be felt because the loss of cognitive function robs people of their ability to live a normal, fulfilling life.

Fear can be felt because this disease has no cure and strikes such a large percent of the population.

Shame can be felt because the medical community hasn't discovered effective treatments and patients can become aggressive and be an embarrassment to their families.

Pity can be felt for families that have been devastated because their loved ones don't recognize them anymore.

Kindness can be felt by the possibility of being able to support efforts to help the millions of Alzheimer's patients.

When used well, pathos moves the audience to feel the same things as the speaker. This creates a bond, or similarity, that can contribute to building trust. Be careful to use emotions that are genuine, though! Feeling an emotional connection makes the listener more compelled to act. Fear is perhaps the easiest emotion to elicit. Take these examples:

The longer it takes to clean up the water, the more people will die from preventable diseases. We must find commercially viable and scalable solutions to slow the death toll, and that's where my research shows promise.

The rapid spread of the Zika virus must be stopped before it gets out of control. We have to invest in the development of a vaccine now and educate people everywhere to mitigate the damage.

Recognizing when your audience is gripped by fear is essential for persuasion. Recall how Bill Young (whom we met in chapter 5, "Why Scientists Communicate Poorly outside Their Field") recounted sending Genentech scientists to the Food and Drug Administration numerous times to give seminars explaining how recombinant DNA technology works.[12]

It took time for Bill's team's safety message to sink in and for it to be apparent that the new technology wasn't going to suddenly cause a problem. You must be empathetic to the emotions of fear and worry and build your case with sufficient logic, proof, and credibility to persuade decision makers to act—and then be patient. This heavily impacted Amy's work in cellular communications when the first concerns around electromagnetic frequency (EMF) from cellular towers became a lightning rod for communities watching out for the well-being of citizens.

In the 1990s, the electromagnetic frequency scare, ignited by a lawsuit claiming that cell phone usage caused thirty-one-year-old Susan Reynard's brain tumor, became part of many public discussions about new cellular applications and decisions about where towers should be located. Amy's employer at the time, AirTouch Cellular, recognized the importance of meeting with local government and community groups. The communications and government-relations teams worked to balance the fear of the technology against the benefits of greater safety and continuous and ample cellular coverage in key areas in case of emergency. They also educated listeners about the fact that signals emanate *away* from a tower. Counterintuitively, it turns out that the lowest amount of EMF is found directly under a tower.

Steve Bohlen gives us an example from when he was at the U.S. Geological Survey of how making an emotional appeal got the rapt attention of a U.S. senator.[13] He also demonstrates the importance of using iconic images and carefully selected language, which you learned about in chapter 8, "The Power of Language." Steve was talking with Washington senator Slade Gorton about building support for the global seismographic network, which was of interest to Gordon's local university, the University of Washington: "I had my fifteen minutes with him and as we got up to leave he looked up at the huge portrait of Mt. Rainier in his office. He asked, 'Should I be worried about that?' And I said, 'Well actually you should. It's a *decade volcano*.' That was the key phrase. And that caught his attention." Steve now had an invitation to continue: "I explained that a decade volcano is on the list of an international volcanological society that feels these are the most dangerous volcanoes on the planet. It was as if fourteen grappling hooks grabbed this guy."

The busy senator, who was very powerful at the time, was eager to hear more, so Steve went on: "I explained that when they erupt, volcanoes always find the same path. I pointed at housing developments in the Deschutes area. I said it's a great place to build houses, nice and flat, easy place to build, good ground water, but sometime in the next 10,000 years or a few hundred years, that neighborhood is toast. He was absolutely a student for an hour."

Steve luckily stumbled upon this prime pathos opportunity. It's well worth giving some thought in advance to how you can use pathos to your persuasive advantage. We are not advocating using scare tactics or inflating the danger of a situation. As scientists, your ability to make your case factually as you address concerns is the ideal approach. That's where logos comes in.

Logos

When you provide sufficient evidence in a well-structured order, you are more likely to convince people to accept your point of view—provided they like and trust you. Cohesive logic is compelling to any decision maker relying on their intellect (System 2 in Kahneman's approach). Incorporating data, examples, statistics, and analogies that help make your case is what constitutes logos. Make your evidence vivid to help bring your science or argument to life. This type of experiential proof is what causes shifts in people's perspectives and enables them to see a situation through the eyes of others who support what you are doing.

INFLUENCE GROWS WITH TIME

Influence is a process, not an event. You have probably had the experience of talking to a colleague about a new idea and having them immediately endorse it. Your professor may have been quickly supportive as well. But you are far less likely to have that experience with your manager or other key decision makers. The newer the idea or the further it is from a decision maker's current activities, the longer it will likely take them to absorb it. You have to allow for this natural ripening of ideas to occur, especially when a paradigm gap exists.

A safe assumption is that it may take multiple interactions with a sponsor just to get to the point where they understand why your idea is important. How should you structure those interactions? Always start with a meeting where you discuss why there is a need. If the decision maker doesn't agree with you on that, continue the discussion and find out just where *they* think the need is. If you can arrange a casual meeting as the first interaction, it will pave the way for the ripening process to occur. You can say something like, "My organization is working on technology X to solve that need, and I would like to come back sometime and give you a briefing on it." Then tell them one reason why your new approach is going to impress them, but don't expect them to be impressed at this meeting. Your goal is to create intrigue and interest.

You can take a similar approach with your manager when you want to seek support for pursuing a new area of research. In that case, the conversation might go something like this: "I've been reviewing the litera-ture and some early findings from our current experiments, and I'm seeing something I didn't expect. It's got me thinking that there may be value in taking a closer look. I'd like to fill you in and get your advice. Can we set up a time for you to come by so I can show you what I'm seeing?"

In that follow-up meeting, ask for their input on your idea. Describe it in a little detail (don't show slides at this point, but perhaps use a drawing on a whiteboard). See if you can find out what they like, what they understand, and what they regard as the actionable part of the problem. Engage them in solving the problem you are jointly examining. *Make them part of the solution.* Even if the decision maker is not at all technical, they may be aware of important issues relevant to your idea. The decision maker could know about a limitation to your approach due to regulatory or ethical concerns, and they may know how to tweak your idea to avoid it. You'll want to conclude this meeting by asking for the opportunity to continue the conversation after you have had a chance to factor their input and the current discussion into your thinking.

The next step is to book a longer meeting where you will ask for the decision maker's buy-in. In this meeting, propose a path forward, either with a few slides (perhaps five!) or a short prepared verbal pitch. It's not time for a formal proposal yet. Listen carefully to the response. Be pre-pared to adapt. If you handled the first series of meetings well, the deci-sion maker will have been thinking about your idea and *placing it in the context of the rest of their work.* This is critical. Now you need to adjust your idea to fit into their world, which has many other needs and priorities. Once you have refined your idea and know that it fits, you are ready to make a formal proposal. Depending on the size of the idea, this first proposal could be a small confidence-building activity, a workshop to ask for other's thoughts on the topic, or a staged effort with go/no-go milestones to maintain confidence. In chapter 16, "High-Impact Exam-ples of Championing for a Cause," we tell the LIGO story, in which almost a decade of lead-in work took place before the first proposal was made to build a working observatory. Hopefully, not all of your efforts will require that much time.

THE PITCH PROCESS FOR INFLUENCING VENTURE CAPITALISTS

If you successfully clear the hurdle of submitting your business plan and getting a venture capital firm interested in an initial meeting, it is vital that you understand what information the firm expects you to communicate each step of the way. Recognize that you are building on their interest and deepening their level of understanding with each interaction you have. VCs tend to commit more time than we just described to their initial discussion with you, but they also place much more importance on it.

Preparing for Your First Meeting

When you have been invited to a first meeting with a VC, you will want to work closely with your contact at the firm. Find out what the participants expect to hear during that meeting and what steps are involved in their overall funding decision process. Nancy Floyd from Nth Power observes, "The good entrepreneurs are the ones who establish a relationship with that partner in the firm and treat them as their champion."

She also notes that in a typical first meeting, you may have only an hour to cover a wide range of topics and demonstrate that you understand the need, the market, the potential customer, and the competition. You will also have to talk through how you plan to use the funding, staff your team, and grow your business. VCs also want to hear how much money you will need to get to the point where the company is mature enough to be bought or go public. Oh, and it's ideal to leave some time to answer questions. So if you have only an hour with a VC, it's clear that you have to extract the essence, and you certainly can't spend too much time explaining technology or discussing your science.

The goal is to get the prospective investor excited and eager to take more time to learn about your idea and its potential. Despite the temptation to get into technical details, remember there will be plenty of time for that during the due diligence process. After the initial meeting, you are likely to have two or three more. These sessions may last a day or more. The firm will come to your lab, office, or garage to see you in action and spend time with you and your team. That's when you can give them a more in-depth look at your science and dive into the details. You will want to reinforce the key messages from the first meeting and bring them to life with demos or evolutionary drawings that help paint a picture of how the firm's investment will enable you to fill a pressing need with an indispensible new solution.

Nth Power won't make an investment decision until everyone at the firm has met the company's entire founding team or management team. The partners report to each other on all active due diligence proceedings to prepare the group to make a decision. All the partners attend these meetings, ask questions, and talk through potential funding arrangements. At Nth Power, a due diligence meeting will usually last for two to three hours. Nancy tells us that a lot is riding on this final meeting for you as a scientist *and* for your contact at the VC firm: "When I spend a lot of time on a deal, I want to win my partners over. I've seen something in a deal that I really like. I've probably spent weeks and weeks working on a deal and I want that deal to go through. There's nothing worse than when one of your own partners turns down your deal. So as a scientist or founder, you are a partner with that champion within the venture firm and you want to make them successful."

Your Team Dynamics Are Part of Persuasion

Nancy also emphasizes how team dynamics influence funding decisions: "Part of what inspires confidence in a team is when a founder. . . lets everyone on the team participate in the presentation." She cautions that a CEO who brings in their team but does all the talking is making a misstep if they are pitching to Nth Power: "We watch the interpersonal dynamics very carefully. How does the team hand off to one another? How do they answer questions? Does everybody defer to one person? If so, is that person too dominant? We may also test the team by asking questions we know will not be well received to see if the team reacts defensively. We may already know the answer but ask the question anyway just to see how the team handles it."

She underscores that Nth Power's goal is to determine whether it can work effectively with the team and whether it trusts the team to deliver on its promises: "Doing a startup is really, really hard and everyone hits a speed bump along the way. I want to know how they are going to pull together as a team when that happens. I may have a unique vantage point about the importance of relationships because I'm a woman but my entire team wants to know what the team dynamic is because that matters." During the due diligence process, Nth Power works to build relationships with each team member. If they do fund a company, they want to be able to go directly to the functional area leads for information, not just the CEO.

The importance of team dynamics is a key part of making the right first impression in many venues. Nancy explains, "I've been a judge for the MIT Energy Prize, for many clean tech competitions and for National Renewable Energy Lab events where you are judging ideas. Even in a 20-minute presentation, we are always looking at what the dynamic is within the team."

Nancy lets us in on a little-known yet vitally important consideration when choosing your team for a venture-funded endeavor: "There is an unspoken rule that the team should not include family members or married partners." She explains that investors in Silicon Valley and in Boston and elsewhere don't want to take that risk. The risk is higher because a wife will not fire a husband, or a father or brother. This makes it harder to get rid of a family member, even if they are not performing. And most investors don't want to take that risk. "And, by the way," she adds, "the risk is even higher with married couples because something could happen to their relationship."

MORE PERSUASION TECHNIQUES AND PRINCIPLES

"It's important for people to understand persuasion for what it is—not convincing and selling but learning and negotiating."[14] This quote is from Jay Conger, the Henry R. Kravis Research Professor in Leadership Studies at Claremont McKenna College, and it illuminates an important dynamic: your ability to champion science depends on how well you listen, understand other points of view, and continue to make the case to convince others of your position by providing compelling stories, facts and data, and emotional appeals.

An excellent approach is to incorporate statements that summarize your own data or data from other recognizable and credible sources:

> The data is perfectly clear. It shows that this heat transfer technology delivers efficiency gains that change the economics significantly. Here are the cost impacts.

> Research compiled by analysts from NASA, as well as organizations from five other nations with space programs, suggests that a moon colony is viable with international support.[15]

Data helps build trust. We found that Josh White, introduced in chapter 9, "Designing Effective Visuals," was very conscious of the importance of striking a balance between helping people understand that they can trust you know your stuff and keeping them engaged at a level that's conceptual enough that they don't feel intimidated, even if they don't know exactly what you are talking about. He astutely observed that the ability to do this is what makes a speaker really effective: "Speakers often focus on the technical content and forget about other cues, like visual and emotional cues. One important element is

Ten Principles of Persuasion for Scientists

Dave Straker operates a website called ChangingMinds.org that is dedicated to principles of persuasion. The following ten, based on Dave's principles, are key for scientists.*

1. **Alignment:** Make sure everything you say lines up so there are no contradictions to cause disagreement.

2. **Amplification:** Emphasize your most important points by drawing more attention to them. Make your voice louder. Pause to let them sink in. Repeat the points you want people to remember.

3. **Consistency:** Maintain consistency between what you say, how you say it, and your body language.

4. **Contrast:** Highlight the difference between where we are today and where your science could potentially take us.

5. **Evidence:** Provide compelling proof and visual examples whenever possible. Vivid stories and examples make a memorable impact.

6. **Framing:** Meaning depends on context, so set and control the context.

7. **Interest:** Interested listeners pay attention. Whet your audience's appetite with a compelling or surprising introduction.

8. **Logic:** Make sure your presentation is rational and cogent. When things make sense, we are more likely to believe they are true.

9. **Repetition:** Reiterate your important points. If you repeat a message often enough, your audience will eventually hear it.

10. **Understanding:** Be clear. If your listeners understand you, then they can interact more accurately with you.

*"Persuasion Principles," ChangingMinds.org, www.changingminds.org/principles/principles.htm.

that you have to give people a feeling of the structure of a presentation to make them feel comfortable as it unfolds. Give them the feeling that they will understand it."[16]

Josh's subtle navigation in the header of his slides (figure 16, chapter 9) is one means of giving your audience a sense of structure without belaboring the outline. If your listeners are feeling lost in a technical section, they can take comfort in knowing which topics are next. Keep in mind that you always run the risk your audience will get overwhelmed with technical details and stop listening. Yes, your listeners will have an emotional reaction to what you are communicating. They may feel confused, curious, unsure, or skeptical, but ideally you want them to be excited about the promise of your science. Your ability to make your ideas clear, understandable, and compelling helps you create the emotional reaction and support you are seeking. But what about your own emotions? How do you come to terms with how *you* may be feeling?

12

Managing Your Emotions

Being a champion of science requires you to be a visible advocate. It demands excellence in your communication skills. If you can't convey your ideas in a clear, convincing, and compelling manner, you might not succeed. How does that make you feel? Nervous? Unsure?

As a student, studying your scientific discipline leaves little room in your schedule for mastering general communications skills, let alone public speaking. Yet once you enter the work world, you have to establish yourself as a credible representative of your science. You have to talk science, look the part, and come across as authentic. How do you come to terms with self-doubt and the other negative emotions that can impact your ability to be an effective champion?

Fortunately, there are several techniques and practices you can learn that, when combined, will ease the distress of communicating about your science.

DEALING WITH STAGE FRIGHT

It's a rare scientist who doesn't get gripped by nerves—especially when they are making a presentation. According to an article by *Forbes* magazine contributor Nick Morgan, when it comes to speaking in public, an estimated 80 percent of people get butterflies and feel anxious.[1] Perhaps they don't sleep well the night before, but they know they will live

Breathing Techniques to Calm Your Nerves

If you need a way to calm down quickly, try some good old deep, slow breathing or learn a breathing technique like the 4-7-8 (or Relaxing Breath) exercise recommended by a noted wellness physician who is at the forefront of preventative medicine, Dr. Andrew Weil. You can find a description on his website (www.drweil.com/health-wellness/body-mind-spirit/stress-anxiety/breathing-three-exercises), and a helpful instructional video on YouTube where he demonstrates the proper technique (www.youtube.com/watch?v=_-C_VNM1Vdo). The alternate nostril breathing technique,* which you can find with a quick online search, also works well to soothe the nerves. For thousands of years, this method of calming the nervous system has been used extensively in Ayurvedic medicine and yoga. Performing it for just a few minutes can quickly reduce stress and fatigue, making it an ideal tool for tamping down speaker anxiety.

*See, for instance, Melissa Eisler and Nadi Shodhana, "How to Practice Alternate Nostril Breathing," The Chopra Center, www.chopra.com/articles/nadi-shodhana-how-to-practice-alternate-nostril-breathing.

through it. It's just not much fun. There's even a term for the fear of public speaking: *glossophobia.*

The classic fear response sends adrenaline pulsing through your nervous system—the proverbial fight or flight reaction. Since you know you can't run, you are stuck with all that extra energy. Trying to contain it doesn't help. It's like trying to make a puppy sit when she would rather chase the squirrel she just spotted. Rather than reining in that burst of adrenaline-powered energy, you can put it to work to make your voice louder and your gestures larger. You can also dissipate some of it by moving purposefully across the room or closer to your audience if you are speaking in a forum with enough space.

If you are going to be in a meeting where you will be sitting still, try to use up some of that extra energy right before you arrive. If you can spend a few minutes in a place where you won't be seen, shake off your adrenaline rush like a swimmer getting ready to race: make big circles with your arms, breathe deeply, bend forward, and stretch your legs. Jog in place a minute or two if you still feel a huge surge of energy that you want to expend.

CONFIDENCE-BUILDING TIPS

Be careful not to psyche yourself out by focusing on how important or prestigious members of your audience are. In most cases, the people you are meeting with are open to hearing what you have to say. They are likely to either be neutral or somewhat favorably inclined toward you. Your invitation to address them is an indication of interest. And no matter what their title or position of power may be, your listeners are fundamentally just fellow human beings. Don't inflate the fear factor.

Enthusiasm Projects Confidence

In her book *Presence,* Amy Cuddy reminds us why it is so important to exude confident enthusiasm: "There's another reason we tend to put our faith in people who project passion, confidence, and enthusiasm: these traits can't easily be faked. When we're feeling brave and confident, our vocal pitch and amplitude are significantly more varied, allowing us to sound expressive and relaxed. When we fearfully hold back—activating the sympathetic nervous system's fight-or-flight response—our vocal cords and diaphragms constrict, strangling our genuine enthusiasm."[2]

Find a Friendly Face

One way to boost your confidence is by making eye contact with attentive people in your audience. When people are following what you are saying, they tend to nod in agreement, gaze in your direction, and maintain a relaxed body position. *These are the people to focus on.* Their level of engagement will help inspire you to give your best delivery. It is human nature to perform better and remain energized when you are getting positive reinforcement from listeners. In small or well-lit venues, watch for supportive listeners—they are often the same people you connected with when you were warming up the room (see "Connect with Your Audience by Warming Up the Room" in chapter 4, "Who's Listening?").

What If You Are an Introvert?

Author Susan Cain, introduced in chapter 4, talks about how she learned to be a confident speaker and provides guidance for introverts like herself in her book *Quiet:* "I've come to embrace the power of the podium. For me this involves taking specific steps, including treating every speech as a creative project, so that when I get ready for the big day, I experience that delving-deep sensation I enjoy so much. I also speak on topics that matter to me deeply, and have found that I feel much more centered when I truly care about my subject."[3] If you watch her TED talk, you will see that you don't have to be an extrovert to be a great speaker.[4]

Conversational Eye Contact Helps Calm Speaker Jitters

One of the most natural ways to make eye contact in a large meeting room or from a podium is to mirror what you would do in a normal conversation with a small group of people. Don't stare. Don't flit your eyes back and forth between people. Don't appear mechanical by moving your gaze all the way around the curve of people in the audience in quick succession like an automatic sprinkler.

Instead, catch the eyes of one person while you say a single sentence, or two at most, and then shift your gaze to another. This is called conversational eye contact. Holding eye contact any longer will start to feel like you are staring. If you are in a large forum and can't see individual faces, you can concentrate on one area and imagine someone looking back at you. Combining physical movement toward your listeners with varying your eye contact is also effective—especially if you are giving a presentation at the front of a room.

The beauty of making conversational eye contact is that it helps you pace your speech effectively and sets up a dynamic that reduces speaker jitters. It makes it seem like you are having a series of individual conversations, which tricks your mind into feeling more comfortable when you are speaking in front of a group. Conversational eye contact also works well when you are seated at a table in a meeting. It helps keep your listeners engaged.

Dressing for Success Still Matters

Last on the list of confidence builders is dressing well. An appropriate outfit (see chapter 6, "The First Two Minutes") will look professional and not distract listeners from hearing your message. What you wear can also impact your performance, according to a study conducted by Northwestern University.[5] The phenomenon, termed *enclothed cognition,* was brought to light in a series of tests conducted to measure the cognitive abilities of people when dressed differently. Enclothed cognition involves the co-occurrence of two independent factors—the symbolic meaning of the clothes and the physical experience of wearing them.

In one scenario, subjects wearing street clothes were tasked with finding differences in a series of pictures. The same test was also given to subjects wearing the type of lab coats worn by medical doctors. When the results were compared, the group wearing lab coats performed substantially better, making half as many errors as the others.

In another case, similar tests were given to subjects who were all wearing the same coats. One group was told they were lab coats. The other was told they were painters' smocks. Again, people who thought they were wearing lab coats performed much better than those who thought they were wearing smocks. The conclusion in both cases, that individuals dressed like doctors focused more and performed better, is notable. If we extrapolate from those results, the notion of dressing for success in clothes that flatter you can only help you deliver a better presentation.

What to Do When the Adrenaline Hits

We often hear that timing is everything in life. With that in mind, give yourself a quick pep talk right before you address a decision maker or audience. Take an "I've got this" mindset into the room. Tell yourself that you are ready to go and will do your best. Remind yourself what you have done to prepare for this moment and why your work and passion will make a difference. Descriptive, specific strength-based reminders help you be yourself in front of your listeners. This kind of positive emotional encouragement also offsets that voice of fear in your head that often fuels a fight-or-flight response.

Your mindset and how you see yourself will affect how you perform. Boosting your self-confidence is the foundation for improving your speaking ability. Gazing at friendly faces, making conversational eye contact, learning how to calm your nerves, making sure you have your message carefully crafted, and leaving ample time to rehearse will help you feel more confident. Try these approaches and figure out what combination works best for you.

AVOIDING SELF-SABOTAGE

Self-doubt can make people vulnerable to self-sabotage. There can be a strong temptation to admit aloud that you are nervous, as if saying it to

---

Combatting Dry Mouth

Do you ever find that you suddenly feel like your tongue is sticking to the roof of your mouth? Adrenaline can do that to some people. Dry mouth can make it hard to talk. The best fix for this predicament is a singer's trick: suck on a lemon wedge. It's sure to get your saliva flowing again. If you have a history of this happening and don't expect to be in a place where you can ask for a glass of water with lemon, bring a slice along in a plastic bag and suck on it just before you start your talk. Amy often does this. The beauty of this simple fix is that it has the added benefit of cleansing your vocal chords, which purifies the sound of your voice. You may also want to avoid consuming dairy products right before you speak because they can coat your vocal chords, which can make your voice sound raspy, or cause you to clear your throat repeatedly, which can annoy listeners.

---

the people you are addressing will make the feeling go away. Voicing your fear in front of decision makers usually has the opposite effect. It amps up your discomfort and may also erode their trust in you. If you want to talk about your fear, do so with a trusted colleague or family member.

Feeling nervous isn't something to fear. It's just your body's way of confirming that the discussion or presentation has high stakes and that it really matters to you. If your listeners see or sense that you are nervous, that's different from you saying it. Most decision makers would expect some sign of nerves. The good news is that even if you feel your nerves are a nine on a scale of ten, your audience will perceive them to be more like a three or four. Amy has tested this over the years while teaching public speaking workshops. Take comfort in it. Besides, the goal isn't to erase your nervousness; it is to keep it from overwhelming you and sabotaging your ability to think straight.

Consider that you are not just nervous—you are also excited. Feeling excited can help unleash your natural enthusiasm. Try telling yourself that you are excited by the opportunity instead of the usual, "I am so nervous." Making that simple mindset shift, along with learning how to capitalize on your nervous energy, can help you quickly improve your effectiveness as a speaker.

Another corollary form of self-sabotage happens when you realize you've skipped a point you meant to make in a talk and exclaim, "I forgot to say . . ." You've just called attention to a flaw. There is no harm in showing some vulnerability, but it can distract you and shake your confidence. Besides, it's not necessary. Remember that your listeners are not mind readers. They will never know you forgot to say something unless you admit it. Instead, just find another opportunity to weave in your point. The missed remark is usually brought to mind by something in the current topic—try to highlight that for the audience by saying something like, "This highlights an important consideration from our earlier discussion." The trick is to quickly move beyond the negative self-talk that arises when you realize you didn't make your intended point. The adage "Keep calm and carry on" is your best defense.

Lastly, don't let your worry become a self-fulfilling prophecy. If you find yourself imagining you will be a dismal failure, you have to change the story you are telling yourself. It's time for some *what if* scenario thinking to anticipate pitfalls and be prepared to address objections. Ask yourself: What is the worst thing that can happen? Where could I get stumped? If the decision maker doesn't immediately show interest, how can I anticipate challenges and prepare for them? What will I do if the slides won't project? How will I handle tough questions? This mental practice helps you come up with a plan of action so you don't have to think on your feet. Preparation and mental rehearsal go a long way toward easing stress.

The goal is to be very present and not get caught up in what Amy Cuddy describes as "second-guessing ourselves and attending to the hamster wheel in our heads—the jumbled, frenetic, self-doubting analysis of what we think is happening in the room. The excruciating self-awareness that we are, most definitely, in a high-pressure situation. And, we're screwing it up. Exactly when we most need to be present, we are least likely to be."[6] Practice, and envisioning your success, help alleviate this fear factor.

VISUALIZATION HELPS YOU FEEL WELL PREPARED

A powerful method for boosting your confidence is to rehearse your conversation or presentation in your head while picturing yourself in front of the people you'll be addressing. Mental rehearsal sets the stage for a more comfortable experience. This technique has been studied and proven extensively with athletics. Research shows that a combination of imagined practice and actual practice often results in better performance than that achieved solely with actual practice. Studies have also proved that imagined practice improves performance in diverse contexts that include communication, education, and clinical and counseling psychology.[7]

Mentally rehearsing the steps in the process, including picturing things that might go wrong and how you plan to recover, will help you be your best and reduce some anxiety. Here is an opportunity to become self-aware. Try to identify what's making you nervous and what else you can do to be better prepared. Now you are ready for a mental rehearsal. In his LeaderLetter newsletter, Wright University management professor Dr. Scott William recommends the following procedure, based on the work of C. C. Manz and C. P. Neck.

1. **Find a Private Place.** Find a time and place where you won't be interrupted.

2. **Recline.** Recline or lie down, and close your eyes.

3. **Relax.** Relax, concentrate, and focus. Take deep breaths and exhale slowly. As you exhale, imagine that stress is leaving your body. Start at your feet . . . feel all the stress leave your feet . . . then your legs . . . then your chest . . . all the way to the top of your head . . . feel all the stress leave your body. Free your mind of distractions and allow yourself to focus on the relaxation process.

4. **Focus.** Once relaxed, focus on the specific challenging task.

5. **Believe.** Mentally tell yourself that you are confident and that you have the ability to perform this task successfully. Repeatedly tell yourself, with confidence, that you will be successful.

6. **Visualize.** Imagine what you will see just before you begin the task. Visualize yourself as an active participant, not as a passive observer.

7. **Rehearse.** Remaining relaxed and focused, mentally rehearse successful steps in the performance of this task. Imagine going through the process and seeing successful results.

8. **Repeat.** Repeat step 7 several times.

9. **Praise.** Finally, open your eyes and smile. You have successfully performed in your mind, which is great preparation for actual performance. You should now feel more confident that you will perform successfully in the real situation. Remember to praise yourself for being successful. Self-reinforcement is another a key to self-motivation.[8]

Conducting a mental rehearsal while out walking in a park or other natural setting can be a great way to bring yourself into a more relaxed state of being where your subconscious mind is better able to absorb the material. For some people, the shower works well, too.

If you try these techniques and still feel overcome by fear and anxiety, there may be something bigger in your way. Maybe you were severely embarrassed or teased when you were speaking in front of a class years ago. Perhaps a very critical coach or parent left you feeling you weren't good enough. If you think a traumatic past experience is amplifying your self-doubt, it's best to identify it, remember what happened, and recognize that you are much more capable as an adult. Seek professional help if you can't shake the negative feelings. Feeling confident and prepared can help you recall your material, especially when the rush of adrenaline hits.

MAKE TIME TO BE PREPARED: CREATING A PRESENTATION MAP

The most impressive speakers do not give impromptu presentations. They invest time in learning the flow of their talk so they are well prepared to deliver it. They develop opening comments that grab attention—whether they are having a conversation with a colleague they bump into in the hallway or giving a formal presentation. They also premeditate how they want to close to be sure they have a clear call to action and reinforce key messages. Even the content in the middle is thoughtfully crafted to move from one idea to the next.

Transition Statements

Mapping out transition statements makes a discussion flow smoothly and prepares champions to use time wisely. Transition statements are what you use to get from topic to topic or slide to slide during a presentation. They help you organize your thoughts in a way that others can follow. The practice of creating and rehearsing transition statements will help ensure your content is in the ideal order.

For presentations, we recommend that you print out your visuals in handout format with three per page and write down the main points you want to deliver for each slide. Also, write out your transition statements so you can see how they enable the flow of the conversation. Once you have done this, give some thought to whether you want to rearrange material to improve the overall flow. Practice the presentation aloud and time it. If you need to cut content, now is the time to work through what you can leave behind. It sure beats doing it on the fly in front of your listeners.

Presentation Map

Once your content and visuals are solid, take your preparation to the next level by printing thumbnail versions of your slides nine-up on a page. You will find this option under the Print Handouts menu in most presentation programs. This configuration will reduce the number of pages you have to study and let you see the order of the slides. More important, it will give you empty space where you can write out notes containing your key messages and transition statements. The act of writing them down helps embed the information in your memory. A printout in this format serves as a visual presentation map to help you learn your materials.

Close with Impact

Never leave your closing comments to chance. This is your opportunity to reinforce your ask, reiterate your key messages, and leave your listeners feeling like their investment of time has been worthwhile. Fumbling for a close or, worse, just ending by saying thank you is for novices, not champions. Plan your closing remarks so you can end as powerfully as you opened.

Finally, don't worry about forgetting some of the elements in this map. Of course there will be some! The point is that you are so much *better* prepared with it than you would have been without it. It's impossible to be perfect. Better is the goal.

TIPS FOR EFFECTIVE PRACTICE

Research shows that the period just before falling asleep is an ideal time to get content locked and loaded in your mind. Saying your material aloud is key if you have an auditory learning style. For best results, use the two or three days before an important meeting or presentation to get the messages and transitions polished, create a visual map, and study the flow and content. Try reviewing your content while you are on the treadmill or pedaling a stationary bicycle. If you are a procrastinator by nature, you'll have to push yourself to get ready in advance so you have a few days to absorb your content fully.

**Advice for Overcoming Stage Fright:
Practice and Coaching**

While it may strike fear in your heart, the more you put yourself out in the public view and communicate, the easier it will become. Anticipating feelings of stage fright and performance anxiety help you come to terms with them. The more you practice and have success, the more you desensitize yourself to the grip of fear. Seeking out opportunities to present inside and outside of your professional circles can help you build up your communication muscles. The Toastmasters organization is a fabulous place to learn new skills and practice what you already know. There also are many affordable and effective workshops on becoming a better public speaker. Look for one where you get videotaped and receive constructive "playback" feedback to help you refine your content and your messages. Personal speaking coaches can also be worth the investment to help you become a true champion.

Practice with an audience of supporters who can listen to you speak, ask questions, and give you constructive suggestions for improvement. This is an opportunity to approach experts who are more senior to you and ask for their advice. They may not have time to hear your entire presentation, but you can go to them to test a few elements and see if they are clear and compelling. You may want to seek their guidance on the best way to respond to tough questions.

Roger always rehearses his talks in his office with the door closed, standing or sitting as he would in the venue, with his presentation map in hand. He doesn't say everything in the talk out loud—just the important opening, closing, and transition statements—but he makes sure to think about what the content is in each slide. That way, he can practice a long talk in a few minutes, honing the most important elements.

IMPOSTOR SYNDROME—A COMMON CONFIDENCE-SHAKING EXPERIENCE

In her book *Presence*, Amy Cuddy discusses the impostor syndrome:

The general feeling that we don't belong—that we've fooled people into thinking we're more competent and talented than we actually are—is not so unusual. Most of us have experienced it, at least to some degree. It's not simple stage fright or performance anxiety; rather, it's the deep and paralyzing belief that we have been given something we didn't earn and don't deserve and that at some point we'll be exposed. Psychologists refer to it as impostor syndrome, the impostor phenomenon, impostor fears, and impostorism.[9]

In many cases, all it takes to start feeling like an impostor is to arrive at graduate school and interact with faculty or advanced students. They know so much about the field and have such amazing insight that you could never imagine matching them. Pair that with the incredible difficulty of graduate classes, and it can be pretty easy to believe that you don't belong and that you could never reach the heights of scientific achievement.

What we don't realize is how hard great minds like Marie Curie, Albert Einstein, and Richard Feynman had to work, how much their signature accomplishments were in fact the summation of many people's efforts, and how they too agonized about their place in the scientific hierarchy. Science is an ecosystem. Each scientist has an important role to play. And just like in any ecosystem, there is always a bigger fish. Every scientist, engineer, doctor, technician, administrator, or artist can look around and say, "Wow, I wish I had the skills and accomplishments of that person. They are truly accomplished." Go ask that person, and they will point you to yet another. Even the most esteemed and accomplished scientists have experienced this. We wish we could ask Curie, Einstein, and Feynman who their bigger fish were. You can bet they all had one.

When Roger arrived at Caltech in the summer of 1978, there were Nobel laureates on every sidewalk, or so it seemed. As a chemistry undergrad who decided to pursue geology in graduate school, he faced the added challenge of tackling a new field at a school where excellence was expected of everyone. Needless to say, he quickly found himself thinking he was an impostor and that his chemistry degree must have fooled the Caltech geologists into thinking he was smart. He was nearly sent back to remedial mapping class after his first day. He was taking engineering mathematics from the author of the leading textbook and quantum mechanics from a Nobel laureate, all while trying to establish real research in a field where he had only the most superficial knowledge. How long could it be before they realized their mistake?

Roger's savior was a new faculty member, Ed Stolper, who went on to become Caltech's ninth provost in 2007. Ed had torn through his undergraduate degree at Harvard, continued to get his master's at the University of Edinburgh, and then returned to Harvard for his PhD, jumping straight to an assistant professorship with his two-page-long publication list at a time when that could be the length of a geologist's entire career of publications. Talk about terrifying the graduate students: as an *undergraduate,* Ed had discovered that a group of meteorites known as Shergottites were from Mars. No spacecraft had yet reached Mars. We knew it only from telescopes. Yet Ed had the insight to realize that that giant volcanoes of Mars erupted under much less gravity than Earth, and so the magmas would have different chemistry as they melted, even if the starting material was the same as Earth. He made the calculations, did the experiments, and darned if the Shergottites weren't an exact match. We had pieces of Mars in our hands.

Clearly, he was a big fish. But over beer with Roger at the downstairs bar at the Athenaeum where graduate students were permitted, Ed admitted that he was pretty sure he was going to be discovered to be an impostor.

Roger vividly recalls this as one of the great moments of release in his life. If this accomplished young professor felt he was an impostor, it clearly was not a function of his achievements. It was a function of the world we live in, and we were all subject to it. Roger knew he could never match Ed's level of accomplishment, but he also knew that the feeling of dread was shared. People rarely talked about it at the time, and few do today, but Roger remains forever grateful to Ed for releasing him from that burden.

Amy Cuddy studies impostorism. Does it surprise you that she was subject to it as well? She says of her early days in graduate school:

> I don't just study impostorism, I experienced it. And I didn't just experience it, I inhabited it. It was like a little house I lived in. Of course, no one else knew I was there. It was my secret. It nearly always is. That's how impostorism gets such a good grip—it pays you hush money. If you don't tell anyone about those feelings, then people are less likely to think, "Hmm . . . maybe she really doesn't deserve to be here." No need to give them any ideas, right?[10]

Cuddy had the experience early in her graduate career of telling a professor that she just couldn't make it—the class required a twenty-minute presentation that she just couldn't see her way to complete. She told the professor she was going to quit rather than face the class.

"No you're not," said the professor. "You are going to go out there and do the talk, and you are going to keep doing it—even if you are faking it—until you realize that you *can* do it." And she did. Cuddy finished the psychology degree, went on to complete a postdoc, and then landed a position on the faculty at Harvard. One day, an undergraduate came into her office and told Cuddy that she could not complete Cuddy's class because 50 percent of the grade was class interaction. Cuddy gave her the same response that her professor had given her years ago. Only then did she realize that *she no longer felt like an impostor:* "And at that moment, it hit me: I no longer feel that way. . . . But I didn't realize those bad old feelings were gone until I heard the words coming out of her mouth. My next thought was this: She's not an impostor, either. She deserves to be here."

Finally, impostor syndrome is not limited to the field of science. One of our favorite examples comes from Neil Gaiman, the author of multiple best-selling novels, who has talked of his feelings of being an impostor. He tells this story of his realization that impostor syndrome is in many famous people's heads:

> Some years ago, I was lucky enough to be invited to a gathering of great and good people: artists and scientists, writers and discoverers of things. And I felt that at any moment they would realize that I didn't qualify to be there, among these people who had really done things.
>
> On my second or third night there, I was standing at the back of the hall, while a musical entertainment happened, and I started talking to a very nice, polite, elderly gentleman about several things, including our shared first name. And then he pointed to the hall of people, and said words to the effect of, "I just look at all these people, and I think, what the heck am I doing here? They've made amazing things. I just went where I was sent."
>
> And I said, "Yes. But you were the first man on the moon. I think that counts for something."[11]

PART FOUR

Applying the Championing Science Skills

13

Translations, Templates, and White Papers

The first thing you learn in media training is, for god's sake, talk in sound bites. Turns out, that's all the training you need to work with interpreters when you are overseas, as I often was with DTRA. You kill your interpreter if you talk too much.

—Jay Davis

During your scientific career, you are likely to encounter several common situations that require special communications considerations. This chapter provides guidance on handling three such challenges so you can address them with confidence.

MULTILINGUAL AUDIENCES AND WORKING IN TRANSLATION

In today's connected world, you will often find yourself talking to audiences who don't speak your native language. This is a time when extracting the essence is critical. (See chapter 3, "Extracting the Essence.") There are a few things to keep in mind.

Speak Slowly and Simplify Language

If you are lucky, as Jay Davis was when traveling for the Defense Threat Reduction Agency, you will have an interpreter. If this is the case, you have two new constraints. First, you have to speak slowly and incorporate significant break points that will allow your interpreter to repeat your message. Even if you are using simultaneous interpretation, where the interpreter attempts to speak at the same time you do, you should give the audience breaks to absorb the translation without the distraction of hearing you speak in another language.

Second, keep your wording simple. When communicating through an interpreter, simple, straightforward language is particularly important. Interpreters rarely have a technical background. While they will generally translate *heart attack* just fine, *myocardial infarction* might occasionally come out as *farm animal*. You should always have (short!) key messages written on your slides.

Interpreters will often request slides in advance so they have time to look up unfamiliar words. They like to be sure they know what slide titles mean. Having the message as the slide title can help the audience too, since many people can read in a second language much more efficiently than they can process speech. As always though, keep those elements short because the audience cannot *listen* and *read* simultaneously, especially in two languages. Speak slowly and pause often. There is no shame in pausing long enough to give your audience time to understand.

Allot Space for Translation

Sometimes your slides will be translated—the second language will be added next to your original. As you create your slides, leave space for this, and stick with very simple charts and figures since they do not tend to get translated. Garr Reynolds gives some great advice in *Presentation Zen* on how to edit slides with written translations.[1] Make the text of the translation smaller so that it does not visually conflict with the original language. It is good practice to point out to any audience what the axes are on a graph, but it is especially important here. Use the laser pointer carefully, and give the description in simple phrases. Make sure to always state what you conclude from the chart.

An interesting note: the direction in which audiences read influences where you should stand relative to the screen. English and European languages read left to right, which means that the audience's eyes return to the left side of the page, or the screen in this case, after the completion of every line or thought. That is where you should be standing so that they are not forced to glance back and forth across the screen to make eye contact with you. Arabic, Japanese, Chinese, and Hebrew speakers will look to the right side of the screen to find you. When you walk into a new room without an established speaker location, use this rule to place yourself.

DEALING WITH PRESENTATION TEMPLATES

"Here is the fifteen-slide template for your fifteen-minute presentation."
At times, a sponsor will mandate a specific template—and they usually
have good reasons for requiring it. Often it's because they have seen too
many presentations that left out important elements. It is common for
templates to be used for competitive pitches or for program review
meetings. You may find that this is true in your own organization, especially if there is a budget meeting designed to make decisions about
where to allocate funding. How do you produce an interesting and
engaging presentation when faced with a list of requirements that is
already longer than the amount of time you have available? Where can
you fit in the material that will set you apart?

Honor the Rules

This trick is simple, and you will find that almost all the successful presenters use it: don't *follow* the rules, *honor* the rules. Provide all the
required material, labeled so it can be easily found, but don't feel compelled to stick to the exact template. Let's look at a few examples where
you can bend the rules and some where you should adhere to them.
Here are suggestions for some of the important common items in most
templates.

Schedule

You may be asked to provide a detailed three-year schedule of tasks and
deliverables, for example. No readable slide can provide this level of
detail. You may be tempted to use three slides to make it understandable—but often, the better answer is to employ a dual format.

In your presentation, make a single slide carefully titled *exactly* as in
the template so that a reviewer can find it easily. On that slide, summarize the work in *one* major deliverable for each year or budget period.
Even though your plan (adequately described in your written proposal
or report, which always accompanies this sort of thing) has more detail,
create a higher-level statement that characterizes the progress you will
(or did) make in the year. That is the level of detail someone wants
when watching your presentation.

If you haven't provided written material, you can provide backup
slides. They are really supporting documentation in slide format. (See

chapter 9, "Designing Effective Visuals.") There is no need to refer to these slides in your presentation, except in the rare case that a question arises. Even then, answer the question verbally and add that the details can be found in the backup material. This answer moves you ahead of your competition from a workability viewpoint. It demonstrates that you can use the sponsor's time effectively, have a good grasp of your own project, and are able to provide details when necessary.

Title

Make sure that you include all the requested information on the title slide because reviewers often have a stack of presentation printouts and need to know which is which. But template title slides are so boring! If you can, use your very best photo on your title slide. Make sure there is room for the text and that it doesn't cover up the good parts of the photo. Typically, the title slide is up on the screen for a little bit before you start speaking, so take the opportunity to get the audience interested right away.

Background, Need, Motivation

When these topics are included in a presentation template, it implies that the sponsor has seen so many bad talks that they need to beg presenters to explain why they are there.

If you do a good job with your first two minutes, there is no need to burden your presentation with slides with these titles—but of course the content must be there. These topics are an intrinsic part of the five slide approach and should be in any presentation when the goal is more than simple education. Put in the material, framed and presented in a way that will excite the audience or at least make them look up from their laptops. But create titles that present the essence of your problem space. *Don't* use the words "background," "need," or "motivation."

Applicability

Be sure to *keep* this title (or one like it) in your slides. Every reviewer who likes your proposal will later be looking to make a salient comment about why they like it, and that comment will almost always be extracted from your material on this slide. Make it easy for them to find and copy into their notes.

But don't be a slave to the format—add your descriptive summary to the title as well. For instance, in a case where you are proposing a new catalyst, a title like "Applicability—This Solves the Need for Faster Process Kinetics" honors the template while giving the reviewers the comfort of knowing exactly what you think the applicability is.

Milestones, Milestone Progress, Gantt Chart

Organizations depend on accountability to justify spending money on science. This is true whether it is the National Science Foundation, a pharmaceutical firm, venture capitalists, or your institution's internal investment committee. The larger the organization, the more important the paperwork.

This can be very frustrating to scientists who are diligently producing results at the quickest possible pace. We will see a great example of this in the LIGO story, where Caltech almost lost the project because of their inattention to the importance of a work breakdown structure (see chapter 16, "High-Impact Examples of Championing for a Cause"). Why is the project sponsor so focused on milestones when they seem to understand your progress and problems perfectly well? The answer is twofold. First, you need to make sure they really do understand your progress and problems. These are never accurately portrayed in a Gantt chart. Take the time to describe the issues, both positive and negative, without hanging some obscuring format on top of them.

Second, the sponsor or reviewer is always accountable to some higher authority who knows relatively little about the overall program and almost nothing about your part of the program. This higher-up has hired a manager or put together a review team with the time and experience to know whether the organization's money will be well spent. The responsible party needs to be able to condense your progress into bite-size chunks and then track those chunks for many projects on a similar basis. Milestone progress is an understandable way to do that. It enables your overseers to defend your work to the next level of oversight. Making that easy for them makes you a good prospect for further funding. Choose milestones that are accurate and achievable. Be honest about them.

Discussing Milestones

How should you deal with the request to discuss milestones in a templated presentation? Summarize, highlight, and provide backup slideuments. For

---

How to Handle Schedule Slips

If you are running behind schedule on your commitments, show a slide that honestly states that and explains how you plan to get back on track. A surprising result, the loss of a key contributor, the failure of a supplier to deliver on time—sponsors have experienced these kinds of setbacks themselves, and a good sponsor will help you come up with solutions. In any case, it is always better to present the problem on your own terms rather than have someone else criticize you in the reviewers' comments.

---

a template slide titled "Milestone Progress," make sure to add an appropriate modifier, such as "On Track." Mention your next milestone or two explicitly on the slide, with expected completion dates (remember that being even slightly early is very positive, but be honest). Then provide a detailed chart or list of milestones as a slideument backup, footnoting its presence on the summary slide. This lets the reviewer know that the data is available but also shows that you did not waste your influence time with them trying to explain an unreadable slide. Use that time to talk about the impact of the next phase of the project, and let the reviewer read the details when you are off the podium.

Combine Template Items

Uncomfortable with leaving out any of the words specified in the template? Most young scientists are, particularly when presenting to a new sponsor. Cheat a little. In this case, due to the typical fifteen-slide topical constraint, it's okay to make an exception to our earlier advice of dedicating a single slide to a single topic. You can combine two or three topics to make one slide title. Budget and schedule often combine nicely. This keeps you from being tempted to provide forty or fifty words for each topic on separate slides, which would take you over the fifteen-slide limit.

We all tend to keep adding detail in hopes of somehow convincing the sponsor that we have everything they need. But what they really want in a presentation is the overview—the reassurance that if they will not be wasting their time if they read your detailed proposal. Keep the details on paper. Keep your slides as stripped to the essence as you can.

Summary

Please do not insult your audience by rehashing what you just said. At the end of your presentation, excite them, give them confidence, or let them know how to take action on what you have discussed, but do not simply summarize. They heard your talk, they have your slides in front of them, and they are *not* stupid, otherwise they would not be listening to you. Respect them. Close with your key messages, not a verbatim repeat. This is especially important when making proposal pitches, which are often templated presentations.

The oft-repeated advice, "Tell them what you are going to tell them, then tell them, then tell them what you told them," is not for science audiences. It is for people who are being forced to listen to you and don't care about your message. Good sponsors are bright and interested in your topic. Don't treat them like mules. A true champion has a different mindset. Tell them why your work is important, what it is, and how they can participate.

Title your last slide with the key message that you want the listeners to take away. This is a point where you can firmly ignore the template title. Turn and look your audience straight in the eye and deliver that message. Leave the slide on the screen while you thank them for their attention and ask for questions. They will understand that was your summary.

Backup Slides—Just Don't

If you create backup slides as written documentation, avoid using them during your presentation. Even if you can't quite remember the detail required to answer a question, resist looking for your backup slide. Instead, simply mention that the details can be found in the slides that you are providing.

Why is it a mistake to show a backup slide? First, it creates a clumsy transition time when you are focused on the computer and not the audience. Even fifteen seconds of distraction wastes the incredibly valuable opportunity to look the questioner in the eye and convey information of value. Second, it makes it seem like you don't understand your own material.

Finally, when you have backup slides prepared, you tend to read that material into any questions that are asked. Listen carefully and answer the question that is actually asked rather than the one you prepared for in your backup slide. Backup slides sap your will to create a lean, focused presentation. When you are going through that initial phase where you have too much material for the time allotted, it is tempting to just dump all the extra information in the backup slides instead of editing it down. The only clear case for backup slides is if you include them as intentional slideuments for the purpose of documenting details to be read later, as discussed above. Don't show them.

Attribution

Attribute all material extracted from other people's scientific presentations. In general, referencing someone else's work in a presentation is considered fair use and is within copyright constraints since you are providing a form of review. Make sure it is adequately referenced. Giving the last name of the author and the publication date—for example, Aines (1987)—is often enough if the name is unusual and the topic clear. Put it in the body of the image—the lower right corner is the conventional spot. Don't ever *print* anyone else's work without explicit permission, however. If you must reproduce a graphic from someone else, make sure you apply your own understanding to it—highlight your point or the learning that helped you move forward. *Analyzing* the work of others is acceptable in most venues. Don't just steal other people's stuff.

Be the Presenter Who Stands Out—Positively

Honor the information needs that the sponsor considered in making the template. Step outside the lines, but pay attention to where the lines are. This gets you credit for being responsive and respectful and lets the sponsor know you understand their world. Stepping slightly outside the template is a simple adaption of the five slide approach (see chapter 3, "Extracting the Essence"). It reassures listeners that you will be a nimble and effective manager of their money. Titling the slides they are likely to need to look at when writing their review (like "Applicability," "Progress," and "Budget") makes it easy for them to praise you.

WHITE PAPERS

Scientists are often asked to explain their topic and outline their ask in the form of a white paper—a short (two pages typically) description of the problem and your proposed solution. White papers are extremely useful when approaching a new organization because they give your contact a simple document to circulate to gauge support. It is a rare decision maker who can act entirely on their own volition—they generally need at least some consensus within their organization.

A white paper is *not* a scientific discourse on a topic. It is a conversation on paper. Don't think of it as just a description of the work you are going to do—think of it as motivation for the work you want to do. *Why* should be highlighted first, and *what* should come next, with approximately equal weight.

Almost none of your readers are likely to be expert in your science. However, they are usually quite familiar with the needs of the topical area. We find it useful to think of the potential audience being characterized as either primarily concept-oriented or primarily detailed-oriented. Detail-oriented people are not great audiences for white papers because they want to see the full proposal. Gaps in your logic or plan will stand out to them—it is critically important to write for them at a uniformly high level without glaring gaps. The second dimension that distinguishes your audience is whether they fundamentally like your concept because of some predisposition or they will never like your approach or goal. This creates the four audience types shown in figure 17. Three of the four deserve your attention.

Scientists often make the mistake of assuming that a white paper is best aimed at the detail-oriented opponents of their idea (lower right quadrant). But you cannot provide enough detail in two pages to change the mind of a nitpicker who thinks your idea is unworthy. You can, however, give the other three audience types enough ammunition to carry your idea forward in spite of the presence of some opposition. Give your strong supporters (upper left quadrant) the chance to defend you, and build support in the remaining two light grey groups that are more likely to be swayed.

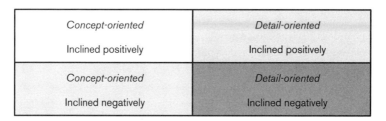

Concept-oriented Inclined positively	Detail-oriented Inclined positively
Concept-oriented Inclined negatively	Detail-oriented Inclined negatively

FIGURE 17. Types of white paper readers.

White Paper Structure and Detail

Use a nested five slide approach to write white papers. Begin with a paragraph that covers the slide topics described in chapter 3, "Extracting the Essence":

1. The problem
2. The technical gap
3. How to fill the gap
4. Why you or your team are right for the job
5. The ask

Expand on each of those sentences in the following five paragraphs. After you finish, you will go back and say, "There isn't enough detail to convince the skeptics!" Look at figure 17 again. Skeptical, detail-oriented readers are a small portion of your audience. Spend your valuable space focusing on the other three phenotypes that are likely to support your proposal. Most scientists write white papers with way too much detail that fail because they do not convince readers that the topic matters.

Constructing White Papers

The first paragraph of a white paper should focus on your concept. In about five sentences, you need to summarize the entire paper. Begin by identifying the problem and the scientific gap associated with it. The gap is more important than the problem. Think of it as the missing knowledge that keeps us from accomplishing things we care about. Then use a sentence or two to discuss how you will fill that gap. Close the first paragraph with why you and your organization are right for the job. Include what you will provide for the money, like a new model or a demonstration of concept.

Written well, this first paragraph will be convincing to the half of the audience that is concept oriented. To solidify their support, you need to add a few details so that they are confident you know what you are doing and can deliver on the big concept. Your next two paragraphs, completing the first page, are expansions on the topics introduced in the first paragraph. You might consider using one very simple figure or table, like the one in figure 17, with a small number of elements. In these two paragraphs, you want to provide one deep dive that demonstrates you really are an expert in your field. Remember, most of the readers are not likely to know much about you. Providing one important detail or referencing a key relationship demonstrates you have a solid grasp of the topic. Giving anything more will make readers skip the paragraph.

The second page of a white paper will get much less attention than the first, so consider carefully whether you need it. If you are responding to a request with a specified format, save the required boilerplate details for the second page (and minimize them). Feel free to leave white space at the end. Spend a paragraph describing why your team is perfect for this task. Multidisciplinary teams always play well with new sponsors, in part because they are surprisingly rare. (See chapter 11, "Influence and Persuasion.") Talk about how your idea sits relative to others in the field, but don't imply that no one else has ever had your idea. That suggests it is likely a poorly poised idea. You want to talk about how you can execute the idea better than others or how you have combined ideas in a unique way. Here are a few important considerations to keep in mind.

Use 12-point font with normal spacing. Be respectful in the construction of your white paper. Don't use a tiny font to squeeze in more detail than anyone has time to read. Remove text if it doesn't fit. Edit your comments so that only the most important ones are present and thereby emphasized.

Clearly indicate the organizations involved. Put the money and time required in an obvious place outside the text. "Three years at $400,000/year" is unambiguous. Describe what you will deliver, as discussed in chapter 3, "Extracting the Essence."

Show your passion. Bring the reader along for the exciting ride that you experience when you think about doing your work. Money tends to go to the people who show they can deliver the work and who indicate that they will take the nucleus of investment and expand it into more exciting and valuable work by inspiring other scientists and funding agencies. Passion counts.

These situations give you a chance to put your championing science skills to use to package your content in formats that are helpful to decision makers. You may have multiple opportunities to share your science in these formats while you are developing trust and rapport with important influencers and decision makers who can advance your work. How do you build relationships with these people over a long period of time and foster shared goals? How do you turn interest into partnership and then into accomplishment? The next chapter answers these questions.

14

Strategies for Creating Successful Relationships with Sponsors

When you are talking to a decision maker, your tasks at the end of a presentation or discussion typically are directed toward helping that person take action. The end of the presentation is the *beginning of a process*. You want to find out how you can meet the needs of the decision maker—then you need to follow through.

MAKE THE ASK: USE A DECISION MAKER'S TIME EFFECTIVELY

Scientists often find it hard to make specific requests. Jay Davis pointed out that when a busy person makes time for you in their schedule, it is because they are interested in what you are doing. If you leave the room without asking for something, you've wasted their time. You almost certainly booked the meeting because you wanted something. Ask for it.

But this is definitely where things get hard. A common reason a scientist is in front of a decision maker is to ask for money to pursue a project. In the conventional agency form of science, there is no uncertainty in making the ask: a sponsor puts out a call for proposals that specifies exactly what they want and how much they are willing to pay to get it done.

When you don't have the benefit of that script, you have to spend more time figuring out just what the decision maker would be interested in supporting.

**When You Get What You Want,
Stop Talking**

You polished your presentation, thought of answers to dozens of difficult questions, and even shined your shoes. At last, you are presenting your idea to the decision maker. After your first three slides, she announces that you are exactly what she needs and has been looking for. But you still have thirty more slides to show!

So how much more should you cover? When listeners are really connecting, particularly in a small group setting, let them set the direction. When you get what you want, stop talking and start listening. Your elegant presentation doesn't have to go to waste: leave a hard copy of your slides with the decision maker, preferably as notes pages (see chapter 9, "Designing Effective Visuals"). Focus on finding out just what is needed and when. The first thing on your mind should be delivering your sponsor's first request and strengthening the relationship.

- How does their organization support science?
- Have they made prior grants or supporting agreements in your area?
- If so, what were the terms and sizes?

It's important to get a sense of the decision maker's history, even when they are a part of your own organization, like your direct manager. It's perfectly appropriate to ask and show your interest in understanding how funding decisions get made. Who is involved? What's the best way to present ideas? Answers to those questions will help you tailor your approach.

When you are dealing with a new potential sponsor in an unstructured arena, it pays to have an ask ready that is *not* for money. Questions that get the sponsor to talk about their thought process are both useful and endearing. "How are you thinking about developing this area?" or, "What do you think the important questions are?" are simple inquiries that can lead to a substantive discussion. It is rare to hit the exact tone, size, and content that an unstructured sponsor is looking for on the first try. Using your ask to develop a better understanding of their needs can lead to a quick turnaround and a successful proposal.

Roger has been party to a number of initial interactions where the sponsor immediately resonated with the topic. Of course, this happens when scientists are careful to understand the sponsor as well as possible ahead of time. If you find yourself in this situation, you definitely want to have an action plan ready—a way to continue the sponsor's interest. With an eager sponsor, new questions emerge: How can your organization best advance the common cause? What skills and background do you have that can be used to resolve major issues or bring the sponsor's vision closer to reality?

In these cases, it is very important not to let the sponsor's ardor for the topic color your common sense about what you can accomplish and what the most useful actions really are. You are the scientist. Being honest about what will move the field forward will build a lasting relationship with a decision maker, whereas taking advantage of a decision maker's interest to push an agenda of your own that is off-axis with what they need can permanently color the relationship. For instance, you may have spent twenty years studying a related field and would like more money to continue that work, but if you know that there is another approach that will serve the sponsor better, take the high road and let them know that up front.

A new relationship with a sponsor or agency has to be built gradually with a series of successes that instill confidence and build your understanding of what the sponsor needs. When thinking about your initial ask with a new decision maker, aim for things that are bite-sized in their world. A sponsor who intends to spend a total of $1 million is not going to trust a new organization with half of that money right off the bat. No matter how good your proposal might be or how good your resume looks, the sponsor has to have proof they can trust you.

An excellent way to start that trust-building process is to have a third party that the sponsor already trusts speak on your behalf. Most decision makers in scientific fields have developed a coterie of trusted advisors who know a lot more about specific topics and performers than they do. Since an important job of these advisors is to understand the capabilities of different organizations, you can often approach the advisors before talking to the decision maker. Invite them out to see your facilities and meet your collaborators. Then, when you are in front of the decision maker, a simple statement like, "We had your chief technology officer out to visit us last month, and this is the process that she was most interested in," is more powerful than any self-assessment you could make.

MAKE THE DELIVERABLE

Sponsors absolutely *do not* want you to exceed the time you agreed to in order to deliver a more perfect product.

Why is this? Surely they are interested in the science being as accurate as possible?

Look at the issue from their perspective. Yes, sponsors want accurate science. But that's what they assume you will deliver! That is why they contracted with you in the first place. You gave them the impression that you would deliver great science, and now it is your responsibility to deliver it *on time,* because decision makers live in a very complex environment where they are judged mostly by things *other* than the scientific quality of their investments. Your report coming in on time and on budget is something concrete that any board of directors or advisory committee can agree on as a measure of success. Don't put your sponsor in the position of having to defend why you were late to the people who control *their* access to money. Make it easy for them to continue funding you. Get your deliverables in on time.

After you turn in one solid deliverable, a sponsor will be better disposed toward giving you a larger piece of the available funding the next time. Often, your work for a new sponsor is exploratory or an entirely new approach. Your first deliverable to that sponsor should not be the ultimate exposition of the new approach with final metrics and values for use. Instead, give yourself and your sponsor an early present and wait until later in the relationship to tackle bigger challenges.

First Deliverables

Make your first deliverable something simple, like a detailed description of the process that you will use to choose the initial experimental approach. A deliverable of this sort is surprisingly valuable to both you and the sponsor. Of course you are going to determine how to conduct your experiments, so why not get credit for doing so? And the sponsor will appreciate having a solid plan to help them understand why you are making the choices you must make along the way, which will make the whole thing understandable and rational. This will show that you are not shooting from the hip but carefully investing the sponsor's money.

Trust Your Sponsor

What about if you have convinced a sponsor to pursue a dramatic new approach with the hope of changing the way the entire field is considered, and you discover a fatal flaw three months into the project? Too often, principal investigators will try to hide flaws like this from the sponsor, hoping that they will somehow discover a way to fix it. This is dangerous.

Good sponsors can be trusted. When you are championing a new scientific approach, your sponsor is unlikely to be didactic or formulaic. They would not have sponsored you if they were. So your best approach is to first quickly develop a plan for how you will deal with the flaw and then present both the flaw and the plan to the sponsor as soon as possible. The sponsor has already invested in you, so they will be glad that you have a plan to address the flaw—it protects their investment. They will have to report to their oversight board as well, and being able to describe both the problem and your proposed fix will almost always satisfy the their concerns. The sponsor understands that science is imperfect and unpredictable—they just want to make sure their money is well spent.

MAKE THE FRIEND

Championing science provides a fascinating filter on people's interests and inclinations. We tend to be passionate about the science we do because it reflects our worldview. Others who share our passion and our worldview can potentially become friends.

The decision makers you interact with are no different. They are interested in many of the same things as you are. As you develop a professional relationship, don't pass up the opportunity to build lasting friendships. Take your sponsor to dinner when they travel to meet you or when you are both attending a meeting away from home. Introduce them to your family. Talk about your favorite hobbies. Ask them how their children are doing in soccer. Do the things you would do with any friend. It feels odd at first because of our ingrained training from school that equates *decision maker* with *teacher* and puts them out of range for this kind of social interaction, but that is a short-sighted way to think about it.

This is particularly important for younger scientists, who will go on to spend decades in a single field, probably working with the same people. The young staffers sitting in the second rank of chairs at your meeting with today's decision maker are in line to become the next decision makers. It's likely that they will be working in parallel with you for the rest of your career. When you have become a senior leader in your laboratory, they will be heading research institutions and government agencies. Take the time to get to know them.

Connections are one of the most valuable things a scientist can have. We tend to build relationships with our scientific peers, and we do a pretty good job of maintaining those connections. Program managers also need to build connections in order to create a network of reliable providers and advisors. This economy of trusted connections is a kind of symbiosis that runs on good results and good relationships.

Your position as a dependable scientist is valuable to decision makers. Once you have established a relationship (on both a scientific and personal level), follow up with them periodically. Especially when you aren't asking for anything, your presence will be welcome. Ask how their plans are evolving. Ask what issues are pressing for them today. Tell them something really cool that your team has done lately. Ask for their advice on an issue that is important to you. There is no better way to get a busy person's attention than to tell them you have a difficult problem that you think they will have insight into.

Continue the gift economy that we first discussed in chapter 11, "Influence and Persuasion." Bring the decision maker new information—tell them what your organization is focused on or what questions your collaborators have been raising. Expect to get questions, and try to answer them right away. Often, decision makers, especially legislators, don't have much time, and they tend to ask questions that they want an answer to in a few hours. Just remember that they do not need, and probably could not use, much detail. A brief answer provided in time to influence an action is infinitely more useful than an elaborate one that arrives after the action has been taken.

Connections are essential if you want to change the world. Remember Julio Friedmann from chapter 5, "Why Scientists Communicate Poorly outside Their Field"? He has experience both as the decision maker and as an influencer of Congress and administration policy. And Julio was a longtime scientist before moving into those roles. His advice reflects his entire background:

> Always start by using your friends. A lot of scientists are the stars of their own heroic epic—the one that is running in the back of

their mind. In their mental image of Act Three, the republic is about to fail, and they are summoned by the Congress to testify about the crisis. They imagine themselves standing in front of the relevant committee, laying out all the myriad technical difficulties. After their testimony the chairman says, "The veil has been lifted from our eyes. How big a check should we write?"

You and I both know that never happens. Instead, it turns out that friends talk to friends, and people talk to people, creating lasting relationships.[1]

15

Exercises for Applying the Championing Science Skills

Are you ready to try out the presentation and communication skills discussed in the first three parts of the book? This chapter contains questions, exercises, and a checklist to help you prepare for an actual situation or a likely future opportunity.

Working through the following seven steps will help you become a competent champion in your own area of science.

THE STEPS TO MASTERY

1. **Know Your Audience:** Conduct a thorough assessment of your listeners.

2. **Craft Your Key Messages:** Create a set of messages and iconic analogies to use when talking to decision makers about your science.

3. **Map Out the First Two Minutes:** Create an attention-grabbing opening and talking points for the first two minutes of a formal presentation.

4. **Create Your Narrative:** Discuss your science using story format to relate your own experiences pursuing your science.

5. **Take the Five Slide Approach:** Create a short presentation about your science using the five slide approach, applying the principles of great design.

6. **Understand Yourself, Your Paradigm, and Your Communications Skills:** Reflect on your mindset and skills as a communicator to become more self-aware and help ensure a decision maker will understand you.

7. **Write Your Speaker Bio:** Write a bio that builds your credibility and could be used in a conference program or by someone introducing you as a speaker.

KNOW YOUR AUDIENCE (CHAPTER 4)

Create a profile of your audience by researching and documenting the answers to these questions:

Who will you be meeting with/speaking to? Who are the key decision makers?

Who influences the decision makers?

What has the decision maker supported in the past?

Who else will be in attendance?

What shows up about the attendees when you conduct a Google search?

What are their publications?

What can you learn from their LinkedIn profiles? Is there a social link (e.g., you went to the same school or share a hobby) that you can use to start a casual conversation?

Do they use social media professionally? If so, what topics do they address?

What point of view do they appear to have about your area of science?

How knowledgeable do they appear to be about your area of science?

Do you know anyone who knows them?

Can you find out if they are introverted or extroverted?

Are there potential religious beliefs or ethical or safety concerns that might impact your listener's paradigm? What information can you provide to narrow the gap between your paradigm and theirs?

How do you need to tailor your approach if the decision maker is a politician? What if they are a venture capitalist?

CRAFTING KEY MESSAGES AND ICONIC ANALOGIES (CHAPTERS 7 AND 8)

Key Messages

Write a set of messages that capture the main and most important ideas that you want listeners to understand, focus on, and remember. Anchor it to the elements in the five slide approach (see chapter 3, "Extracting the Essence"). Develop these two or three ideas in several short, easy to understand phrases that capture the essence of your science and will resonate with your listeners.

Prepare examples, stories, and data you can use as proof points to support each of the messages you create.

What would the user or beneficiary of your science value most about it?

What action do you want listeners or decision makers to take as a result of your discussion or presentation? Write down your ask.

Iconic Analogies

List the key concepts associated with your science that may be new to listeners or hard to understand.

Is there a parallel to something most people are familiar with that you could use to help them understand these new or complex concepts? What is your equivalent of a heat-trapping blanket or a stuck accelerator pedal? Create a brief explanation of a complex component of your science using your analogy, or incorporate it into your messaging.

The following example illustrates some of the key considerations for establishing overarching messages and support points.

Messaging Example: Preparing to Deploy
Ammonia-Based Refrigeration

In 2013, a major employer with a large campus of buildings and several thousand employees was updating its cooling and heating systems and preparing to start the permitting process with the local government. After weighing the options, they decided to use ammonia for refrigeration.

Background

There are a number of environmental concerns with gases used for refrigeration. Until recently, the chlorinated and fluorinated versions were major contributors to the destruction of the ozone layer. Those chemicals were replaced with versions that don't attack ozone as part of the 1987 Montreal Protocol, one of the most successful environmental treaties of all time. However, the replacement options still contribute significantly to global warming. The release of these gases represents 3 percent of the total worldwide warming emissions.

The good news is that ammonia has proven to be an extremely energy-efficient replacement that does not contribute to either ozone depletion or global warming, and it has been used safely in large industrial systems since the earliest days of refrigeration. The bad news is that it can be hazardous to humans in significant quantities, making leaks dangerous.

The engineering organization at the company was tasked with creating messaging that took into account both the global climate benefits and the potential local hazards of using ammonia for refrigeration. Further complicating the situation was the fact that the community had experienced an ammonia leak at a nearby facility a few years earlier.

Messages

This example showcases the difficulty of championing this important climate benefit while addressing local safety concerns. Message 1 and the proof points highlight the rationale for the value of using ammonia-based refrigeration and position the company as one that cares about the environment. Messages 2 and 3 recognize the potential safety concerns and fear associated with the use of this chemical and address the topic head-on by discussing the company's engineering and containment approach as well as making a distinction between the company and the neighboring facility that had a leak. This message shows the company cares about its employees and the local community and is taking design precautions to ensure safety.

MESSAGE I: COMMITTED TO GREEN AND
SMART BUSINESS

Proof Points

1. We're investing in new infrastructure to save energy and address environmental and climate concerns. We plan to replace ozone-depleting substances with ammonia, a commercially proven,

environmentally clean refrigerant that meets our cooling needs without contributing to climate change or depleting the ozone layer.

2. Our new infrastructure and efficient energy practices will remove 4,600 tons of ozone-depleting refrigerants and reduce greenhouse gas emissions.

3. This voluntary effort demonstrates our commitment to create a greener workplace.

MESSAGE 2: ENGINEERED FOR SAFETY

Proof Points

1. The new cooling system will be built with state-of-the-art, redundant design and operational safeguards to ensure proper operation.

2. Ammonia has been used as a refrigerant for more than a century in a variety of environments, including dairies, wineries, ice rinks, and food and beverage businesses.

3. All ammonia will be fully contained in a segregated area within a building structure. Only chilled water, not ammonia, will be pumped outside the containment structures.

MESSAGE 3: OUR SYSTEM IS VERY DIFFERENT FROM THE NEIGHBORING FACILITY THAT LEAKED

Proof Points

1. Unlike the neighboring facility, our system design centralizes and contains all ammonia in a confined structure.

2. Unlike the neighboring facility, our system will pump only chilled water, not ammonia, outside the containment structures.

THE FIRST TWO MINUTES (CHAPTER 6)

Write an opening line that captures your listeners' attention and gets them to care. Here are six attention-grabbing ways to start a presentation or discussion. Apply several of the following approaches to craft a few opening line options for your talk or discussion. Choose words that will help create contagious enthusiasm for your topic.

What interesting findings could you reveal?

What would your "What if you could . . . " question be?

What can you say to connect your science to you in a personal way that shows your passion and reason for pursuing your science?

What can you predict about the future that is relevant to your area of science?

What can you compare to the past that is relevant to where you are taking your science?

What prototype, sample or demonstration can you show to introduce your science?

Choose one of the opening lines you have created and use it to start writing out your talking points for the first two minutes of your presentation. Remember to frame the topic you plan to cover without giving too much away so you can create an appetite for listening to your talk.

Be sure your talking points show your commitment to your work and what kind of impact you plan to make if the decision maker supports your request. Articulate the value of your work to humanity and society or to your field of endeavor. Be sure to incorporate language that illuminates your expertise. Perhaps you can find a story that illustrates your capabilities to help build your credibility without coming across as boastful. Develop it according to the prompts in the next section, "Create Your Narrative."

The content of your first two minutes needs to set the stage for what you plan to communicate. It should address the following:

What problem do we face today that needs a solution? What is the current solution and why does it fall short? How will you solve it? Why does your science matter?

What details can you leave out so that you keep your listener focused on the high-level concepts? What do you want your listeners to do as a result of hearing your messages?

For guidance, refer to the five slide overview in chapter 3, "Extracting the Essence." You may also want to watch the Three Minute Thesis presentations referenced below to see excellent examples of how speakers have addressed these questions and get an idea of what you want to accomplish in your first two minutes. Write down the things you like about their approaches so you can emulate them.

Michelle Brown, "Disrupting the Domino Effect of Child Maltreatment," College of Education and Human Development, University of Minnesota, 2016 (www.youtube.com/watch?v= Elt_oP8_iZE)

Emily Johnston, "Mosquito Research: Saving Lives with Pantyhose and Paperclips," School of Pharmacy and Medical Sciences, University of South Australia, 2014 (www.youtube.com/watch?v= dhopJdgY6Lc)

Jenna Butler, "Computing a Cure for Cancer," Faculty of Science, Western University, Canada, 2015 (www.youtube.com/watch?v= GydZ1cRKWLg)

Shruti Menon, "Non-Invasive and Easy Diagnosis of Chlamydial Infertility," Faculty of Health, Queensland University of Technology, 2014 (www.youtube.com/watch?v=7BJ11axTqX0)

Maria Gibbs, "Dancing Bridges, a Human-Centered Approach to Prevent Flutter in Foot Bridges," Department of Civil and Environmental Engineering and Earth Sciences, Notre Dame University, 2016 (https://vimeo.com/160787172)

CREATE YOUR NARRATIVE (CHAPTER 7)

It takes practice and some forethought to develop a compelling narrative that will help you build support for your science. Reflect on the challenges and problems you and your team have faced and how you have approached these issues. Can you recall a pivotal time where you learned something vital to advancing your work? Identify that moment and use it as the basis for answering the questions below to develop your story.

What is the setting for your story? Describe it using sensory detail so a listener could imagine what it looks like.

Who are the people involved? They become your main characters.

What do you want the listener to learn or conclude from hearing your story? Does your story reinforce your key messages? How can you refine your story to ensure it clearly supports your messages?

Another method you can use to incorporate story structure into your presentation is to fill in the blanks in your narrative by using Randy Olson's And-But-Therefore model (see chapter 7, "Crafting Key Messages and Narratives"): "This and this are true, but this flies in the face of them, and therefore the following things must be done."

TAKE THE FIVE SLIDE APPROACH (CHAPTER 3)

Create a short visual presentation that follows the five slide format described in chapter 3, "Extracting the Essence." The questions below will prompt you to think about what you need to convey in the first five minutes of a discussion with a decision maker. Build on the content you created in "The First Two Minutes" section and develop visuals to accompany the words.

1. **Problem:** What is missing from the world that you and your listeners care deeply about? What is the overall context for the problem?

2. **Technical Gap:** What is the specific thing we can't do today because of a gap in knowledge or ability? What is the gap that keeps the decision maker from achieving their goals?

3. **How to Fill the Gap:** How can you help your listeners imagine the future? How can you improve the general state of science and technology? Find an image or a series of images that illustrate what the future could look like.

4. **Why You or Your Team Are Right for the Job:** Why are you or your team poised to be efficient and effective? Why is the job well suited to your skills, reputation, and facilities? What have you and your collaborators done previously in this area? A photo of team members in a lab or other work setting can be ideal for this topic. Labeling them by name and area of expertise is helpful as long as the viewer can easily see faces and names. If you are collaborating with several institutions, a visual that mixes logos with names and headshots is a great graphical way to address this topic.

5. **The Ask:** This slide captures what you need from the decision maker. It's the part of the presentation where you enlist your listeners as collaborators, funders, or other support resources. This slide should spell out how much funding you are seeking and describe what you will you deliver in return for the funding.

UNDERSTAND YOURSELF, YOUR PARADIGM, AND YOUR COMMUNICATIONS SKILLS (CHAPTERS 2, 5, AND 10)

This mastery step will help you become more self-aware and identify skill-improvement areas. Write out your answers to the questions below and consider steps you can take to communicate more effectively with people who have decision-making authority.

These questions will help you find the words to convey what makes you passionate about your science. You can also use this insight to help craft the first two minutes of any talk.

Why did you choose to pursue your area of science? Was it because something happened to someone you care about?

What gets you excited about your work?

What makes you feel the greatest sense of challenge and accomplishment?

How would you like to make your mark on the science world? What would you like to be known for?

The next set of questions will help you identify your scientific paradigm, jargon, and other potential barriers that you must address before you can bridge communications gaps that might exist between you and a nonexpert decision maker.

What is your scientific paradigm? What school of learning do you identify with?

What are the beliefs and principles that color your understanding of the scientific area you study?

Are there common religious or ethical considerations related to your science that may color the way a decision maker thinks? Think of specific decision makers you know, if possible.

What jargon do you and your colleagues commonly use?

What acronyms do you and your colleagues commonly use? Which ones should you spell out at all times? Which one or two do you want your listeners to learn?

The following questions and the speaking skills self-assessment will help you become more self-aware and improve your communications and presentation skills. If you are not sure how to answer the questions, you may find it helpful to ask a trusted colleague or family member.

How do you see yourself professionally?

What words would you use to describe yourself?

How accomplished do you feel?

What is your academic or career success story, and how do you live it?

TABLE 3   SPEAKING SKILLS SELF-ASSESSMENT

Rate Yourself: Speaking Skills Self-Assessment				

On a scale of 1 to 5 (1: huge need to improve; 2: considerable need to improve; 3: some need to improve; 4: slight need to improve; 5: no need to improve), rate your need to improve in the following areas:

Voice

Incorporating vocal variety	5	4	3	2	1
Making my voice audible	5	4	3	2	1
Speaking at the proper pace	5	4	3	2	1
Eliminating nonwords, e.g., *uh, um*	5	4	3	2	1
Pausing for effect	5	4	3	2	1

Visuals

Covering one topic per slide	5	4	3	2	1
Eliminating unnecessary clutter	5	4	3	2	1
Limiting the number of words	5	4	3	2	1
Leaving sufficient white space	5	4	3	2	1
Using color purposefully	5	4	3	2	1

Physical

Matching words with facial expression	5	4	3	2	1
Making conversational eye contact	5	4	3	2	1
Moving with purpose	5	4	3	2	1
Standing tall or sitting straight	5	4	3	2	1
Smiling at appropriate moments	5	4	3	2	1
Gesturing to reinforce a point	5	4	3	2	1
Calming speaker jitters	5	4	3	2	1

How much ego do you bring to the way you communicate?

Are you humble, or do you act like you know it all?

Are you willing to listen to others' ideas, or do you just push your own?

Are you able to truly listen during conversations, or are you always thinking about what you plan to say next?

Do you really listen and seek to understand?

How do people react to the way you share information?

Are you an introvert or an extrovert?

Do you take time to reflect on how you come across and how others respond to you?

TABLE 3 *(continued)*

Content					
Developing messages	5	4	3	2	1
Using the five slide format	5	4	3	2	1
Delivering consistent messages	5	4	3	2	1
Incorporating examples	5	4	3	2	1
Telling compelling stories	5	4	3	2	1
Using iconic metaphors/analogies	5	4	3	2	1
Eliminating jargon	5	4	3	2	1
Providing the right level of detail	5	4	3	2	1
Persuasion					
Demonstrating passion	5	4	3	2	1
Establishing credibility	5	4	3	2	1
Making an emotional connection	5	4	3	2	1
Making a logical argument	5	4	3	2	1
Building rapport with listeners	5	4	3	2	1
Earning listeners' trust	5	4	3	2	1
Outside the Talk					
Researching your audience	5	4	3	2	1
Warming up the room	5	4	3	2	1
Hearing the intent of questions	5	4	3	2	1
Deferring response to tough questions	5	4	3	2	1
Providing succinct replies to questions	5	4	3	2	1
Engaging all listeners while responding	5	4	3	2	1

SPEAKING SKILLS SELF-ASSESSMENT

Take the self-assessment in table 3 and identify opportunities to self-correct. You may find it helpful to record a video of yourself delivering a message intended for a decision maker or to role-play the situation with a trusted colleague. Seeing yourself in action provides data that will help you rate yourself.

Once you have taken an honest look at your current abilities, pick an improvement category to work on. Make it a focus of your next few conversations or group discussions. When you reach the point where that component is becoming easier, move on to your next area for improvement. We don't recommend that you tackle multiple categories at once, as that can be overwhelming.

WRITE YOUR SPEAKER BIO (CHAPTER 11)

The goal of creating a speaker bio is to build your credibility. You will want to include information that showcases your expertise for a specific audience. This bio could appear in writing in a program for a conference, or it could be used by a host to introduce you before you give a talk (see the two examples below). Write your speaker bio the way you would like it to sound when someone reads it aloud; this will help you keep the language more conversational than formal. Your answers to the questions below will give you the right content elements to work with to craft your bio. Remember that a speaker bio should be at most a couple of minutes long—extract the essence!

What impact are you passionate about making in the world?

How many years have you worked in the field?

What is the focus of your current scientific work?

What are your major scientific credits?

What other areas of science have you worked on in school or in your career?

What are your educational credentials?

Do you have publications or patents that the audience should know about? What about awards or other forms of meaningful recognition?

What other facts will help establish your credibility? Is there a fun fact that would be entertaining to include?

Sample Speaker Bio for a Lay Technical Audience

For the past fifteen years, Dr. Roger Aines has been building coalitions to develop new climate and energy technologies to manage or repurpose atmospheric CO_2. Today he brings together government, industry, and academic partners to find ways to reduce CO_2 by capturing it or creating products from it that drive economic growth while protecting the planet. Roger is the chief scientist for the Energy Program at Lawrence Livermore National Lab, where he's worked since he left Caltech thirty years ago. Roger's work has ranged from stochastic computational methods to nuclear waste disposal to the geochemistry of carbon capture and storage. You'll often find him at events like this, sharing his knowledge about industrial carbon capture and storage, better use of biofuels, accelerating natural systems that absorb CO_2, and carbon utilization.

Sample Conference Program Bio

For the past fifteen years, Dr. Roger Aines has been building coalitions to develop climate and energy technologies that remove or repurpose atmospheric CO_2. He brings together government, industry, and academic partners to pioneer new ways to reduce atmospheric CO_2 or use it to drive economic growth while protecting the planet. Roger joined Lawrence Livermore National Laboratory more than thirty years ago after graduating from Caltech with a PhD in geochemistry. Over the past three decades, his work has run the gamut from nuclear waste disposal to environmental remediation. He has applied stochastic methods to inversion and data fusion, managed carbon emissions, and developed monitoring and verification methods for CO_2 underground storage. Roger holds twenty-two patents and has authored more than one hundred publications on topics ranging from mineralogy to environmental cleanup to carbon capture and storage. He frequently speaks at industry events worldwide about industrial carbon capture and storage, better use of biofuels, accelerating natural systems that absorb CO_2, and carbon utilization. Roger currently serves as chief scientist for the Energy Program at Lawrence Livermore National Laboratory.

THE CHAMPIONING SCIENCE PRESENTATION PREPARATION CHECKLIST

Getting ready for a big presentation? That's the ideal time to review the checklist (see table 4). Disciplining yourself to carefully develop your content and allowing yourself enough time for methodical practice always pays off. If you are well prepared you will be more confident when you speak.

Remember that passion for your science creates contagious enthusiasm and that the best champions are people who can be themselves in front of an audience. Believing in yourself and being well prepared are key to your success!

TABLE 4   THE CHAMPIONING SCIENCE PRESENTATION PREPARATION CHECKLIST

Activity	Completed
Identify your target audience.	
Research important members of your audience.	
Find out what the decision maker might be interested in supporting.	
Develop an attention-grabbing opening line.	
Develop talking points for first two minutes, including a way to show your passion and why your science matters.	
Develop iconic analogies to help explain key concepts.	
Develop two to three overarching key messages.	
Develop stories, facts, and data proof points to support your key messages.	
Test your key messages with colleagues who can serve as a proxy for the decision maker.	
Develop the five slide content (the problem, the technical gap, how to fill the gap, why your team is right for the job, the ask) and supporting visuals.	
Develop a story or anecdote to incorporate into your talk.	
Assess the whitespace, number of words, use of color, font, and design of your slides for simplicity and impact.	
Develop talking points for each slide.	
Develop transition statements so you can move smoothly from slide to slide.	
Tailor your conversation points or slides based on what you have learned about your listeners.	
Develop a clear call to action and incorporate it into a strong close.	
Print your slides out as handouts, with nine slides per page, and write in your talking points, key messages, and transitions so you have a visual map of your talk to study.	
Practice your talk on your own and record it so you can watch and time it.	
Watch your talk to see how well your facial expressions, gestures, and words match.	
Identify places to pause and use vocal variety to enhance your delivery.	
Practice your talk with a friendly audience and ask for feedback and help brainstorming a set of possible questions.	
Develop succinct answers to potential questions.	
Two to three days prior to your talk, review your visual presentation map at night just before you fall asleep.	
Decide how to dress to ensure you make a great first impression and your attire does not distract listeners.	
Apply tips for managing your emotions to remain calm, focused, and present.	

High-Impact Examples of Championing for a Cause

Championing science is never an event—it is a process. We have taken you through the basics of how good communication, influence, and emotional intelligence skills can lead to success. Here are three examples of the how these principles have been used to change the world.

VANNEVAR BUSH AND THE ONE-PAGE MEMO THAT CREATED MODERN SCIENCE

On June 12, 1940, a few days before Paris fell to the Nazis, Vannevar Bush placed a one-page memo in President Roosevelt's hands. That memo, calling for the creation of a new agency to foster civilian research on military topics, set the tone for big science as we know it today. After a fifteen-minute meeting, Bush had Roosevelt's agreement to create the National Defense Research Committee.

This was the most far-reaching act of championing science that we know of.[1] The new agency would change the government's approach to acquiring science for military use, moving away from the prewar method of research being conducted entirely by employees of the Army and Navy to a new model. Now, researchers would remain in their home institutions or corporations and work under contract to the government. The National Defense Research Committee and its immediate successor, the Office of Scientific Research and Development, sponsored the development of the most significant U.S. weapons of World War II: radar, the proximity fuse, and the atomic bomb.

G. Pascal Zachary's fascinating biography of Bush lays out his highly influential career in delightful detail, including the extraordinary report on the future of science, *Science, The Endless Frontier,* which Bush wrote at Roosevelt's request at the end of the war.[2] The report served as the framework for military research and the formation of the National Science Foundation. In it, Bush wrote, "Basic research leads to new knowledge. It provides scientific capital. It creates the fund from which the practical applications of knowledge must be drawn. New products and new processes do not appear full-grown. They are founded on new principles and new conceptions, which in turn are painstakingly developed by research in the purest realms of science!"[3]

For the purposes of this book, we are more interested in the way he achieved his extraordinary accomplishments than in the accomplishments themselves. Did a one-page memo and a fifteen-minute conversation with the president create science in the form we know it today? Of course not!

The memo was a careful summation of the important points, but the president had been briefed by his advisors on the nature of those points, which had already been extensively negotiated with key influencers and aides. It was a classic embodiment of championing science—several years of careful preparation, adjusting the message, building support, influence, and carefully understanding the context in which the new agency must live. All of that extracted into six bullet points, capped by a handwritten, "FDR-OK," that gave Bush the authority to proceed.

Let's take a look at the six points in the memo and examine the championing science principles they reflect.[4]

"1) [The NDRC would be] attached to the
National Defense Commission."

Every great idea needs a home. Decision makers need to figure out how your work will fit into the hierarchy of the existing structures that govern their world. By selecting the National Defense Commission, an agency that didn't really exist anymore (a relic of World War I, it was composed of the cabinet secretaries of the Departments of War, the Navy, Agriculture, the Interior, Commerce, and Labor, and in effect was disbanded in 1921), Bush enabled Roosevelt to act. He provided a brilliant and official excuse to create a new organization that would influence the highest levels of government without the involvement of an oversight organization. Bush wanted to report directly to the president, and throughout World War II, he did.

"2) Composed of chairman, members from War, Navy, Commerce, National Academy of Sciences, plus several distinguished scientists or engineers, all to serve without remuneration."

Bush knew that he could not go it alone. He needed an oversight board that the government would trust. To achieve his desired civilian control of new military research, he would need to maintain very close ties with the existing powers in this area—the Departments of War and the Navy. Bush believed that the Departments of War and the Navy were incapable of handling their own affairs in the new age of technical warfare.[5] But hard experience had told him that the armed services had a formidable ability to impede progress. He had to have buy-in at Cabinet level, reinforced by the president, for any research the new agency undertook.

Bush gained a keen appreciation of the importance of buy-in, and how its absence could thwart progress, from his previous dealings with the National Academy of Sciences. He had a long and frustrating relationship with the Academy dating back to World War I. Robert Millikan, from chapter 1, "Becoming a Champion," had created a component of the Academy, the National Research Council (NRC), during that war as a means of conducting research in military matters. As a twenty-seven-year-old faculty member at Tufts University, Bush had traveled to Washington, DC, to meet with Millikan to pitch his idea for a magnetic field detector to address the fact that German U-boats owned the Atlantic and U.S. warships couldn't detect them until they were within two hundred yards. Millikan was encouraging even though there were dozens of other researchers studying the problem.

But Bush was skeptical of the red tape associated with the new organization and instead asked Millikan to introduce him to the investment banker J.P. Morgan. Morgan agreed to fund the magnetic field detector work, which enabled Bush to proceed rapidly in the absence of bureaucratic overhead. However, despite technical progress, it did not turn into a useful device. Bush was so far outside the Navy systems that he could not garner their support. When he was finally able to test the device on friendly submarines, it worked. But the Navy refused to field it on the wooden ships Bush required and demanded that he redesign it for steel destroyers. It never found a single German submarine. Bush learned that civilian research had to be integrated into the military systems, although he squarely assigned the error to the Navy, not himself.[6]

As World War II threatened in 1936, Bush helped create, and then led, the NRC's Division of Engineering and Industrial Research, with the explicit intent of using it as a vehicle to further military research. But by 1939 he resigned, frustrated with the Academy's refusal to instigate work on defense topics. Frank Jewett, although a friend and supporter of Bush, led the Academy with the attitude that he did not want it to become "just another agency of the government." In words that surely resonate to this day, he stated, "The Academy was in the position of a doctor waiting for clients; it could not adopt the attitude of an aggressive salesman and initiate attacks on what it regarded to be important military problems."[7]

But the Academy was the home of America's great science, and Bush knew that no science-based political activity could proceed without its acquiescence, if not support. Despite his frustration, Bush knew he needed the Academy. It would have a place on the new council.

"3) Function, to correlate and support scientific research on mechanisms and devices of warfare (except in fields covered by National Advisory Committee for Aeronautics). Concerned with research rather than industrial development or manufacture."

Bush understood that this new organization needed to take a "What if you could?" look at how research might be done at an entirely different scale. At a time when the German Luftwaffe had 4,100 airplanes and the United States Army had 800, there was a major concern about how to manufacture the huge quantity of aircraft needed to match such a superpower. But Bush was not concerned about how to create more of what we already had—he knew that the coming war would be won with new weapons, not more of the old ones. He wanted the new organization to focus on research.

"4) Supplied with funds for office staff, and for financing research in laboratories of educational and scientific institutions or industry."

Everything up to this point has been good management and politics, but this bullet was the first breath of wind from the storm that would sweep through U.S. science, and later the world, entirely changing the relationship between research and government. Bush extracted the essence of a complicated change in the way the U.S. government obtained scientific support into a single sentence. The proposed independent agency, reporting in practice only to the president, would purchase research directly from scientists who remained in their home institutions and companies. Up to this point, the typical method of integrating a university researcher during wartime was to offer him a commission in the Army or Navy. The U.S. Armed Forces conducted the research they needed in house, at institutions like the Naval Research Laboratory, with poorly paid civilian staff. At the onset of World War I, the American Chemical Society offered its help; the Secretary of War demurred, saying his department already had a chemist.[8]

Today, it is hard to imagine a world without government grants to support even the most applied science work in weapons and defense. Bush recognized that there was an untapped resource and that an entirely new approach was needed to effectively eliminate the cultural barriers that kept academic and industrial science separate from the military. This new method would greatly reduce the paperwork and, to a degree, the oversight associated with government research. Most importantly, it would allow the institutions and businesses involved to wholeheartedly support the effort because it helped them grow their own research foundations through new capital equipment, dedicated researchers, and students.

"5) To aid and supplement, not to replace, activities of War and Navy departments."

Bush had never been happy with the research conducted inside the military, but he knew better than to make them an enemy from the start. This memo was the culmination of months of careful construction of political support. He used the now expert staff of the National Advisory Committee on Aeronautics (which provided guidance and research on aircraft development and had been Bush's first major appointment in Washington) to draft legislation. He had the self-assuredness to present himself to the president as the nation's leading scientist in military matters, but he knew that he needed more than ego to support the proposal.

As the new head of the Carnegie Institute of Washington, Bush had broad access to the influential elite of the nation, but he was decidedly more conservative than the Roosevelt administration. Bush was a long-time critic of the New Deal, so it was hard to imagine him being able to influence Roosevelt. Bush sought out Frederic Delano, Roosevelt's uncle. Delano was a railroad executive, city planner, scientific dilettante (he studied Mayan ruins, among other things), and general supporter of technology in the midst of the New Deal, whose interest in industrial planning and purchasing power verged on socialism. He was the influencer Bush needed to convince Roosevelt, and Delano was very receptive to Bush's memo. Bush and Delano met in early May.[9] On May 10, the Nazis invaded the Netherlands on the way to France.

On May 25, as the unstoppable Blitzkrieg rolled through the Low Countries, Delano's request to meet with his nephew Roosevelt was readily agreed to. The first step was the usual staff meeting with Harry Hopkins, the president's closest aid. A former social worker and idealist, Hopkins made an odd pair with Bush, but they immediately hit it off. At a time of intense national need, two intelligent men immediately recognized how their joint actions could address a problem. We can imagine that there was no talk of politics. This is a classic example of bringing the right collaborators and influencers to the project, with the breadth of the team providing the decision maker with confidence. The connection to Delano helped Bush quickly establish trust and credibility.

"6) An Army and Navy officer detailed to work with the chairman."

Bush knew that the research had to be germane to military needs and that the best path to accessing up-to-date information was to work directly with the armed forces. He also knew that information would be better received if it was passed to the military by their own officers than if it came directly from civilians.

An incredibly brief summary of two years of work by Bush, this memo is also notable for what it didn't contain. Where was the money? Bush trusted that Roosevelt would make that right, and he did, although the reformation of the agency into the Office of Scientific Research and Development a year later was principally to formalize the monetary relationship to enable Congress to fund the office by line item. Where were the names of the other seven members of the committee? It is possible that Roosevelt and Bush discussed bringing on key members from MIT, Harvard, Caltech, and Bell Labs, which Bush quickly did, but the names did not appear in writing.

Perhaps most importantly, where is the list of research to be done? Knowing Bush's ego, we can expect that he had a clear idea of what he thought needed to be done, but he withheld those opinions until the Army and Navy had been queried about their research needs. Within months of Bush's meeting with Roosevelt, the National Defense Research Committee had five divisions and thirty-four sections dedicated to topics like nitrocellulose, proximity fuses, and hydraulic fluid. But Bush did not present such a list to the president at the outset, although it is easy to imagine that his discussion with Harry Hopkins touched on this diversity. Bush had the details at the ready and had built the staff and influencer support.

The Bush memo is a masterful example of the importance of enabling the decision maker to act and using influence to build support for the decision. The next example underscores the value of building effective partnerships and ensuring that you provide the information needed to earn and keep the trust of your sponsors.

THE HUNT FOR GRAVITY WAVES

In chapter 6, "The First Two Minutes," we referenced David Reitze's 2016 statement as a great example of how to start a presentation with passion: "Ladies and gentlemen, we have detected gravitational waves. We did it."[10] This simple proclamation is also the culmination of a remarkable exercise in championing science. The two four-kilometer-long interferometers used to make this measurement—one in Hanford, Washington, and one in Livingston, Louisiana—are not remarkably expensive by the standards of high-energy physics today. In total, over about thirty years, the National Science Foundation invested $620 million in their design and construction, including $200 million for the original construction and the same amount for a later upgrade.[11] In an era of billion-dollar atom smashers, this seems like the sort of thing that big government science does.

But the Laser Interferometer Gravity-Wave Observatory, known as LIGO (a classic case of an acronym turned into a jargon noun; see chapter 5, "Why Scientists Communicate Poorly outside Their Field"), was a new kind of endeavor for the National Science Foundation. The cyclotrons and accelerators that we associate with big science were owned by another agency, the U.S. Department of Energy. The National Science Foundation (NSF) funded much smaller "principal investigator" projects. When the first serious proposal for an operating pair of facilities came in at $70 million in 1983, it was by far the largest project the National Science Foundation had ever considered.[12]

There was no way to make it smaller—years of experiments with successive prototypes at MIT and Caltech, at one-meter and then forty-meter scale, had conclusively demonstrated that four kilometers was the required size. A pair was required to conclusively demonstrate that the detection of a gravity wave was not a fluke—detection at one site should be followed by an equivalent detection at the other, offset by the speed of light as the gravity wave passed through the earth. This was the requirement for an observatory: a rudimentary, stripped-to-the-essence way to detect and measure stellar events on the scale of the collapse of two black holes or neutron stars into each other.

The history of this project dates back to the mid-1960s, and you can read the original accounts of it recorded for the Caltech Oral History Project or as a much better *story* in Janna Levin's terrific book *Black Hole Blues*.[13] But we are interested here in the championing of the project. And this process has three fascinating aspects: the (possibly shotgun) marriage between competitors MIT and Caltech that was required to get the sponsor, the National Science Foundation, to approve; the intense desire of the sponsor to see the project successfully completed despite enormous problems with the participants; and the difficulty many scientists who were beholden to the National Science Foundation had with the enormous amounts of cash being poured into this project, to the possible detriment of their work. Three constituencies (the investigators, the sponsor, and fellow scientists) with dramatically different views were all ultimately trying to deal with Congress to get the money to move science forward.

The Marriage of Caltech and MIT

The two titans of tech were intense competitors in gravity wave detection. MIT's effort, older but smaller, was led by Rainer Weiss, who was an outstanding experimenter but not much inclined toward bureaucracy or large-scale science. He was a bit of a tinkerer. In the early stages of his research, he was known for leaving things unfinished or unpublished. But his skills at building exquisite experimental devices were unparalleled. When faced with teaching a class in general relativity although he was fundamentally an experimentalist, Weiss approached the course with as much of a focus on experiments as he could. The students appreciated this as a novel aspect of an almost purely theoretical subject.[14]

Weiss posed a homework problem to the students: How could you detect gravity waves? And because he had a lot of experimental experience with optical interferometers, he led the conversation in that direction, eventually convincing himself that if you had two beams of light at right angles to each other in space, reflecting off mirrors, you could detect the change in the length of the beams as a gravity wave passed by measuring the degree to which the beams interfered with each other when recombined at the right angle source. Having established the concept, he set about building such an instrument, not in space but in an old wooden building at MIT left over from World War II research known as the Plywood Palace.[15]

Meanwhile, at Caltech, theoretician Kip Thorne was interested in gravity waves and astrophysical sources, and he had a graduate student working on a review paper on the topic.[16] Thorne and Weiss met at a NASA meeting on applications of space for gravitation.[17] Forced to share a hotel room, they instantly bonded on the topic of measurements of gravitation, staying up all night charting options. Thorne decided that Caltech needed an experimental presence in the field. He attempted to recruit Weiss, which would have been the end of our interesting marriage-of-institutes story. But Caltech didn't find Weiss's resume to be of sufficient mass due to his lack of publications. Thorne wrote back to Weiss, "There must be pages missing, aren't there?"[18]

Instead, Caltech hired Ron Drever from the University of Glasgow— another brilliant experimentalist, but someone who very much wanted to succeed on his own. Now the race was on at the two institutes. The National Science Foundation was taking an interest in the topic, in no small part because the foundation's new head of gravitational physics, Richard Isaacson, was a theoretician with some background in gravity waves (he had done calculations and believed it was possible to observe them). Isaacson thought there was great science here: "All the advice of all the responsible groups says this is one of the great opportunities in science. Gravity waves will give us unequivocal proof that black holes exist. That's the only way we could possibly get that information."[19] By 1976, the two institutes were going forward collegially but completely independently.

Weiss obtained an NSF grant to examine just how sensitive a detector would have to be and what kinds of siting and operating conditions it would have to have. Caltech struck back by investing $500,000 of its own in a laboratory for Drever to build a prototype forty-meter detector, enormously larger than Weiss's one-and-a-half-meter experiment. "That really got the NSF to pay attention," said Isaacson.[20]

Although not deliberate actions by the institutes, and probably not by the NSF either at that point, these are the two kinds of activities required to launch a major field with confidence that the investment will be worthwhile. The sensitivity study tells the sponsor just how far current science is from obtaining the prize (in this case, detecting gravity waves, which were theoretically well understood at the time), and the larger-scale experiment exposes the difficulties and unknown, as yet, hindrances to making an instrument function. Equations are always more predictable than hardware. Before massive investments are made, both the theory *and* the hardware have to be well constrained.

At this point, it sounds like a marriage between the two institutes would be the most natural thing in the world, but Drever proved not to be the marrying sort. Weiss and Thorne talked extensively about collaborating, and Weiss agreed to Thorne's suggestion to make a two-institute proposal.[21] But then Weiss finally met Drever at a general relativity meeting in Perugia, Italy. In discussions the two had at the hotel there, Drever was adamantly opposed to the idea. According to Weiss's account, Drever said, "I didn't come to Caltech to work with you. I want to do my own thing. Why do I have to work with you?"[22]

That was the beginning of a long and difficult period between the two organizations. They combined efforts on a presentation to the National Science Foundation on the results of the MIT study, which found that a bare-bones project was a $70 million endeavor. The institutes were expecting to compete for the right to complete the project as sole actors.

But the NSF was having none of it. They insisted that a project of this scale needed the expertise of both institutes, which of course Weiss had been courting all along by bringing Thorne and Drever into the conversation. Weiss also faced issues at MIT about support from the administration. Famously, in one meeting between the National Science Foundation, Caltech, and MIT administrators to discuss the possible merger of efforts, John Deutch, MIT's dean of science (and later the director of the CIA), was asked how much money MIT would put into the effort. Weiss was hoping MIT would pay for a project manager, but by his account, "John Deutch asked me for a piece of paper. I don't know if you know John. He's very tall. He grabs the piece of paper, pulls out his pen, and he writes a great big zero and shoves it under [National Science Foundation manager] Marcel's face. He said, 'That's what we're going to do.' And he walks out."[23]

At this point, Weiss was almost a decade into this effort, and his home organization had undercut him. Drever was arguing furiously for his own team, and had attempted to get an independent siting and sensitivity study funded. The thought of a Nobel Prize was already in people's heads. Weiss needed Caltech, the National Science Foundation needed both institutes, and Caltech wanted to go it alone. In the end, the money won. Isaacson of the National Science Foundation pushed the *joint* effort forward.[24]

Was this the shotgun wedding that some called it?[25] We don't think so, because while it's true that the funding agency pressed for the marriage, the true champion of the effort, Weiss, had been angling for it all along. He knew his organization did not have the desire or the nimbleness of Caltech to push such a big project forward. And his own collaborators at MIT were few. So he chose to support the path that made the sponsor most comfortable: instead of deciding between competitors, include the best of both of them. Caltech would lead the effort, but MIT would participate. Weiss had an equal role to Caltech in the oversight and science management.

The Vested Sponsor

The National Science Foundation did not regard this as just another proposal. Isaacson had a personal interest in the area, and he convinced the head of the agency's physics division, Marcel Bardon, that the project could be a signature investment for the foundation. Nobel Prizes were in the offing. And the topical area, gravity, did not overlap with the purview of the other big investor in physics, the Department of Energy. The National Science Foundation thought they could make their mark here. MIT's proposed format—a multi-university consortium with scientific contributions from many players—fit the National Science Foundation's preferred model in the mid-1980s (funding that supported breakthrough science in multiple congressional districts). They knew full well that Congress was not going to write the foundation a new line item if it wasn't politically stabilized by having the support of multiple members of Congress. In Weiss's words:

> The whole idea was something that [the NSF] wanted, and it was quite clearly being orchestrated by Marcel and Rich. Every move of the way was being orchestrated within the NSF, and they were just hoping like hell that we would behave ourselves and do the right thing. And if we didn't, they would try to coach us. But they were very sensitive about that. They didn't try to tell us how to propose. But they would grease the way in certain directions, and hopefully we would follow those directions. And this is when I began to realize that the NSF wanted this something terrible.[26]

Weiss had done his champion job very well. The sponsor understood the value, and Weiss helped the middle management (Rich Isaacson) make the case to upper management (Marcel Bardon). The value proposition was well established. And this is an extraordinarily good thing for those of us who are impressed that they ultimately succeeded, because it turned out that the Caltech team was so wrapped up in the beauty and difficulty of the physics that it repeatedly failed to understand that the sponsor also needed successes in areas *other* than science.

In one surprising event, on the eve of starting construction of the now $200 million project, the National Science Foundation sent a letter to Caltech to stop work on the project. Charles Peck, the chairman of the Caltech physics division, remembers the team being stunned by receiving that letter in early January of 1994:

> There was a little table . . . where all the powwows took place among LIGO management—that is, the three chief people: Robbie [Vogt], Stan Whitcomb, and Bill Althouse. And I saw them sitting around the table, and they looked like they were at a seance. They looked as though something terrible had happened, or as though they were anticipating something terrible. It was clear from the way they were sitting there. There was not a word being said; they were all just sitting there staring at a piece of paper in the middle of the table. So I walked in, somewhat innocently, and said, "What's going on?" And Robbie said, "Have you seen this?"
>
> "No."
>
> So I look at it, and I see that it is a letter from NSF saying, "The NSF does not approve the expenditure of funds to begin building the LIGO site in [Hanford,] Washington." That was to be the first site to be started, and it had to be started before some date—maybe the beginning of April or something—because a certain kind of bird lays its eggs in April and you can't disturb the birds for environmental reasons. At any rate, NSF was not going to provide the funds for doing this until NSF was happy with the LIGO management. "And would Caltech please be so kind as to send the management back to Washington [DC] on January 17th"—I think it was about a week—"for a meeting at the NSF offices."[27]

Congress had just cut the budget for LIGO by $8 million, citing the National Science Foundation's concerns with Caltech's management of the project.[28] The foundation was clearly concerned and angry about Caltech's failure to take into account the formalities of management that a project of this size required to be successful in Washington. Caltech was pursuing the project as a skunk works—they felt that they knew what needed to be done. But the National Science Foundation, and Congress, needed more assurance than that. They needed to know that the money would be well spent and that the project would succeed. Peck continues in his interview:

> It had to do with the fact that NSF had been trying for a long time to get something called a management plan. I had read a letter from [NSF's] Berley sometime earlier, and it was complete gobbledygook to me. And part of it was something having to do with a management-breakdown list. I can't remember the exact name, but it's well known to anyone who's involved in modern project management. You make a management-breakdown structure. I can't remember the exact term, but it's a list of all the various activities that are involved and how they break down into various categories. You know: "Build a tunnel." Well, what do you need to build a tunnel? You take care of this, you take care of that, each task has its own set of subtasks. So you end up with 7.3.2.7.8—as an item—and then a word that describes what that item is and how much it costs and what its time schedules are. That was what we were being asked for, but I hadn't the faintest idea what was being asked for from reading the language of the letter. Robbie didn't either, because Robbie was running this whole project very much like the skunk works, the famous Lockheed group that built the U-2. And that's how he put it himself; he was running it like the skunk works. That was his view of how to do things—cheap, fast on your feet, don't worry about all the complicated schedules and details, just get the job done right and do it cheap and right.[29]

Just like any science manager, the NSF had to report to a higher authority and justify their investments in terms that the Congress understood: money and progress. Caltech had dramatically failed to understand the reporting needs of their deeply vested sponsor. The team was so wrapped up in the difficulty of the physics to be accomplished that it had regarded these paperwork and management activities as annoying distractions. It took the intervention of the president of Caltech and the appointment of a new project director at Caltech, Barry Barish (fresh from managing part of the superconducting supercollider project), to put the effort back on track. The previous director, Vogt, had pushed the project through very difficult times, including when the National

Science Foundation refused to site the project in the preferred location in Maine and insisted on it being moved to Louisiana. But at the fateful January meeting, Vogt lost his temper repeatedly, reinforcing the foundation's impression that he was unstable and unreliable. His replacement, Barish, would ultimately share the 2017 Nobel Prize in Physics with Weiss and Thorne.

At the point of this winter crisis meeting, it is easy to imagine that the National Science Foundation would just cancel the project. Very little of the $200 million had been spent—no dirt had been moved at either site. They clearly had lost confidence in Caltech's project management. Why didn't they just bail? It was because they were deeply vested in the project. They had sold a significant portion of their reputation to Congress in convincing them that the project was worthy of the biggest line item in the history of the National Science Foundation. They had spent years in committee meetings, report preparation, and proposal review. They were not a disinterested sponsor. They wanted this project to succeed. The championing had been well done up to this point. The sponsor had become the champion to the next level of management, the U.S. Congress.

Caltech's dramatic, and painful, changes in management and staffing ultimately delivered the success they needed. The two interferometers began operation in August of 2002.

The Competition

Years of ardent support of the concept had finally yielded success—the National Science Foundation obtained $200 million from Congress, and the project was moving forward. But all the participants knew that the first version of the facilities had a low chance of detecting gravitational waves from the sorts of events they expected to occasionally occur, such as the collapse of two neutron stars into each other. The detectors were just not sensitive enough, and the events would have to be too close to the Earth (well, hundreds of light years, but in space that is small) for them to occur very often. An upgrade was needed, and by the time the operations began in 2002, there was a full-fledged effort to upgrade the system to eight times the sensitivity at the staggering cost of *another* $200 million.

At this point, the alternatives for the money really became an issue, even though there was never another serious, large-scale option for detecting gravity waves. The National Science Foundation had funded a number of small efforts, but no one considered them competitive scientifically with the interferometers in terms of really being able to turn them into an observatory, a place where routine measurements of distant gravitational events could be made. But that does not mean that the champions for gravity wave measurements did not have a vocal set of competitors: all the *other* science that could have been done with that amount of money.

The alternative of spending the money on something else is always far from the minds of scientists pushing a new idea. They develop their case, analyze the benefits and risks, and determine the cost. Then they take their idea to a sponsor, firmly convinced that they have the most cost-effective solution to the problem and that the sponsor will act accordingly.

Unfortunately, there are always alternative uses for the money. In this case, opposition to LIGO came from two sources. The physics community was concerned that this gigantic project would siphon money from their much smaller projects. This was even a concern at Caltech, where LIGO had potentially become so important to the institute that it could not be allowed to fail, even at the expense of other efforts and faculty. In 1995, while LIGO was undergoing these massive management changes, a mentor of Roger's, Lee Silver, said in an interview for the Caltech Archives Oral History Project, "I'm concerned by what the fact that we have become so dependent on a large project like LIGO means financially to the Institute. If we face a future where we must recruit large projects, then this institution isn't going to be what it has been. We will not succeed in attracting some of the best minds that I'd like to see us get. We'll attract some of the best science promoters instead."[30]

The champion always faces this opposition, and rightly so. Although wealthy Pasadena residents financed Caltech's first major investments, by the end of the twentieth century, the National Science Foundation and other principal-investigator-based organizations became the mainstay of funding. Letting the "institutional" project become the enemy of the individual proposer was clearly a danger, and one for which there is no clear solution even today. The two paths are always in competition with each other. To a major extent, LIGO did not generate as much opposition as it might have because it worked inside the mechanisms of the National Science Foundation and passed many peer reviews along the way. The science champion cannot become the science huckster— the quality must be there.

But this was not the only type of competition. Other major scientific efforts also needed money. Steve Bohlen (whom we first met in chapter 4, "Who's Listening?") was running the National Science Foundation's ocean drilling program in 2000, just as the first LIGO phase was starting operations. That program uses a massive "drillship"—basically an oil-drilling platform on a ship—to reach the bottom of the deepest ocean floor and then drill into the rock beneath, recovering core samples of the rock (just like an apple core, but each is thousands of feet long). This gives insight into the deep crust and mantle of the earth, the part that drives volcanoes and moves the giant plates around, resulting in earthquakes and the slow evolution of the face of the planet. Deep science indeed, and expensive. Operating the drillship costs $100,000 *per day*. And configuring the laboratories and facilities is also very expensive.

At the same time that LIGO was looking for $200 million to upgrade their detectors to the sensitivity that ultimately observed gravity waves and allowed David Reitz to make the statement that began this section, the U.S. drillship *JOIDES Resolution* had been operating for over fifteen years. But new science needs and normal wear and tear required a $130 million upgrade to keep the science going. This was not funded by the physics division that was funding LIGO, but Steve still had to compete with the specter of the LIGO scientists, who were asking for a similar amount of money also from the National Science Foundation. Steve says, "We had been collecting samples and doing great science for a decade and had a big user community and established scientific programs. The LIGO team had not yet detected a single event that LIGO was built for, after hundreds of millions of investment, and they needed another $200 million just to have a good chance of making a detection. All the while, *we* knew exactly what we needed to keep the science coming in. It was very frustrating."[31]

Both projects ultimately got their money, and both are achieving their goals today. But the reality that there is always another way the sponsor can spend the money must be in the champion's head. You have to make the case that not only is your science important but also that it makes sense in the context of the rest of the sponsor's responsibilities. The gravity-wave scientists struggled with a huge lack of self-awareness. They had the science, but more than science was needed. Fortunately, other team members brought awareness and correction into the picture, or this story may have ended as a failure, not a resounding success.

MEDICAL APPLICATIONS OF ACCELERATOR MASS SPECTROMETRY

Our gregarious Texan Jay Davis from chapter 3, "Extracting the Essence," has championed more than one major science activity, but his favorite is accelerator mass spectrometry. His experience in applying that new technology in medicine is a great example of how to be self-correcting not just in your approach but also in the way you identify the best collaborator or sponsor. We talked with Jay about his efforts to make carbon-14 (the rare and slightly radioactive isotope of carbon) measurements part of medical research.

Jay was very focused on collaborators from the start. He convinced the regents of the University of California to invest $250,000, in addition to the money Lawrence Livermore Lab was putting up, to fund the startup of the accelerator mass spectrometer facility: "I talked the Livermore Laboratory into signing a memorandum of agreement that said, in exchange for that investment, the faculty and staff of the University were guaranteed access to the facility. So if someone said, 'Why is there someone from China in your weapons lab? And why is there someone there at midnight?' we had it covered. We had a big leg up on getting collaborators from all over."[32]

But the question was, who should they collaborate with? Jay's championing of the facility itself had been fairly easy. He convinced the laboratory to provide the time and a simple building. He also convinced them to purchase a used tandem accelerator that could be turned into a high-powered mass spectrometer:

> I had a good pitch. I said, "90% of life is carbon and the next 5% is hydrogen. I can measure and track the carbon with carbon-14 and the hydrogen with tritium, H-3, hydrogen with two extra neutrons, at sensitivities a factor of a million higher than any technique you have." You don't need to know a lot more than that. That's enough to investigate this topic for its extraordinary promise. I can take an isotope you routinely use in your laboratory and find it a factor of a million more sensitively, and by the way, that means I can freely dose human beings with it at levels we can easily measure. I can freely do human experiments. If I give you carbon-14 tagged aspirin, you are in more danger from the aspirin itself than you are from the radioactive carbon-14. That was a pretty powerful argument.

Jay was right about having a good pitch. He summarized the gap and made clear what was possible with a device one million times more sensitive than any similar existing measurement tool. Then he connected it to dosing humans, which helped get people to care about his science. He also understood the potential fears around the radioactivity and was quickly able to allay them by referencing aspirin since his target was medicine and biological research.

Jay had identified that this work was something that could be done only at a national lab. There were absolutely no rewards for doing it at a university. A physics department would not be interested in medicine. A medical department would have said, "Holy cow, an accelerator!" Jay continues: "The first machine we built filled 7,000 square feet. The timing was unique. It was just when Lawrence Livermore Lab started an internal, long-range funding program. The original criteria for this laboratory-directed research and development was to cross disciplinary lines."

So medicine and isotope science (which Lawrence Livermore Lab had always been interested in) was a reasonable mix. But medicine, and its principal sponsor, the National Institutes of Health (NIH), was terra incognita for the physicists of the lab. The NIH is an enormous funding source, but those funds are focused on principal investigators and institutions that have established track records. Livermore, even more so than MIT in our last example, needed collaborators that the sponsor trusted. And the first thing they had to do, their first championing role, was to convince those collaborators that they wanted to use accelerators in biology and medicine:

> That was tricky and it had a funny backspin. In the lab we demonstrated the technique could work. Like most scientists, we thought the biomedical research community would just fall all over themselves to get at this tool. We thought it would be the most wonderful thing in the world because we could effectively give it away free for a while due to our developmental funding. But it turned out that the barrier with biological research people was very high. They already were staffed with scientists and already funded by National Institutes of Health. And their answer was, "Oh god, we'll have to get new post-docs and learn some new lab techniques and write a grant that's different from the one we have been getting renewed regularly."
>
> From the medical perspective, you have got to have MDs at the lab to touch a human body. So from day one, if you wanted to work in that world you had to be a good collaborator, because our lab had no physicians.

And that worked out just fine for Jay, because the people who were most receptive to the new tools with one million times more sensitivity turned out not to be the laboratory scientists but the medical researchers,

MDs, who worked with patients every day. "They were engaged with sick people and they were pissed off that they didn't have diagnostic tools. They turned out to be the ones who grabbed what we offered by the scruff of the neck. They pulled it forward to the point that the National Institutes of Health would finally accept a proposal, and so help me God, they funded an accelerator inside a weapons lab for biomedical research!"

This is a triumphant championing science story. On November 12, 2013, the Center for Accelerator Mass Spectrometry defended the fourth cycle of their National Institutes of Health proposal for about $8 million over a five-year period. And the National Institutes of Health is funding another small accelerator to go into that lab. This means they will have funded the first and second biomed machines—primarily because of pull from the clinical community.

The key for Jay was recognizing the path he needed to take to champion the technology, from demonstrating extraordinary sensitivity, to finding problems where that sensitivity really mattered in medicine, to locating the researchers who *needed* that collaboration to advance their own goals. Self-aware and self-correcting in message, *and* in the target for that message.

These three stories bring to life many of the principles that science champions must take into account as they build support for their efforts. It's clear that patience is more than a virtue—it is essential. The larger the project, the longer it will take to build support.

A FINAL WORD

The opportunity to help scientists become champions inspired us to write this book because we believe that transformative breakthroughs in science come in incremental steps. Take, for example, Michael Faraday's creation of the first hand-cranked electrical generators in the 1830s. About forty years later, the Belgian inventor Zénobe-Théophile Gramme improved on Faraday's design to create an electrical motor. By 1878, the American Charles Brush was selling massive outdoor arc-lighting systems across the nation, leading to lighting breakthroughs by Edison, Tesla, and Westinghouse that ultimately electrified society.[33] When one new idea gets properly introduced, it paves the way for more discoveries. We've seen this happen countless times. But what if that first idea never got any traction? What would the world have missed?

Go forth and champion your science!

The Champion's Bookshelf

We stand on the shoulders of giants. Here are some of the great books that can provide much more detail on the topics that we have summarized. Many are fundamentally oriented toward business users but contain true gems of knowledge that can benefit scientists and engineers as well. Each reference contributed significantly to our writing—we recommend these books highly.

PRESENTATION BOOKS

Garr Reynolds. *Presentation Zen: Simple Ideas on Presentation Design and Delivery.* 2nd ed. Berkeley, CA: New Riders Press, 2012.

> Reynold's book may be aimed at business people, but it is clear that he is a scientist at heart. His advice for clear thinking and presentation, focused on key topics, will help every scientist become a better communicator and presenter. Don't let the Zen imagery dissuade you—
> this is a book that will make you much more effective as a champion.

Nancy Duarte. *Slide:ology: The Art and Science of Creating Great Presentations.* Sebastopol, CA: O'Reilly Media, 2008.

> Duarte is one of the best slide designers on the planet. Her work for Al Gore stands out as a shining example of making complicated science straightforward and easy to absorb. Some images from that project are available on her company's website, www.duarte.com/portfolio/al-gore. *Slide:ology* often dives much deeper than scientists can absorb, but it is great to know what real artists are capable of. When you are working with professional designers, you will know what to ask for.

Edward Tufte. *The Visual Display of Quantitative Information.* 2nd ed. Cheshire, CT: Graphics Press, 2001.

———. *Envisioning Information.* Cheshire, CT: Graphics Press, 1990.

———. *Visual Explanations: Images and Quantities, Evidence and Narrative.* Cheshire, CT: Graphics Press, 1997.

Tufte is a master at designing beautiful and effective graphics. Absorb these three books and you will give your audiences the gift of being able to rapidly understand your ideas and data. You can apply Tufte's principles to any graph or image you create, whether it's a figure in *Science* or a drawing on a napkin.

Felice C. Frankel. *Envisioning Science: The Design and Craft of the Science Image.* Cambridge, MA: MIT Press, 2002.

Felice C. Frankel and Angela DePace. *Visual Strategies: A Practical Guide to Graphics for Scientists and Engineers.* New Haven, CT: Yale University Press, 2012.

Photography has gone from being a "nice to have" element to being a key part of documenting every aspect of science and experimentation. These two books will help you make your photographs, and the slides and graphics you create from them, into extremely high-bandwidth conduits of information. They will also be stunningly beautiful. Felice is an MIT researcher and the only practicing scientist author of any of our recommended books.

Susan Weinschenk. *100 Things Every Presenter Needs to Know about People.* Berkeley, CA: New Riders, 2012.

Reading Weinschenk's brief summary of one hundred key ideas is an easy way to get background on why the most effective communication methods work. Although each section is very brief, scientists will appreciate the CliffsNotes approach to topics that other authors make you suffer for.

Nathan Yau. *Visualize This: The FlowingData Guide to Design, Visualization, and Statistics.* Indianapolis, IN: Wiley 2011.

There are many books on data visualization and graphing. Yau's stands out because it gives details of how to prepare much more complicated graphs than your typical desktop application provides. When you have a particularly complex data set or want to show intricate relationships, Nathan may be able to help.

COMMUNICATION BOOKS

William Strunk and E.B. White. *The Elements of Style.* 4th ed. Needham Heights, MA: Allyn and Bacon, 2000.

Read this book. Regularly. It will keep you from making mistakes of grammar, usage, and composition that distract your audience. Stick with this older edition that is still readily available. Newer additions are longer, more expensive, and less useful. These later versions with additions by other authors do not follow Strunk and White's signature advice: *omit needless words.*

Steven Pinker. *The Sense of Style: The Thinking Person's Guide to Writing in the 21st Century.* Reprint. New York: Penguin, 2015.

Pinker gives scientists two useful guides: how to write with the subject in the foreground and the author in the background and, most importantly, how to think of your writing (and speaking) as a camera that turns the listener's view toward the subject you want them to observe. He gives the only cogent explanation we've seen of why the passive or active voice is the right choice for your situation.

Daniel Kahneman. *Thinking, Fast and Slow.* New York: Farrar, Straus and Giroux, 2011.

Understanding *why* your audience absorbs information, or does not, is essential for a scientist seeking to communicate outside their peer group. Kahneman's fast-and-slow thinking model is an outstanding way to think about how people understand new information and make decisions based on it.

Susan Cain. *Quiet: The Power of Introverts in a World That Can't Stop Talking.* New York: Crown Publishing Group, 2012.

Extroverts are not the only great communicators, nor are all decision makers extroverts. Cain helps us understand how introverts make decisions and how those of us who are introverts can better get our ideas across to others.

John Maeda. *The Laws of Simplicity: Design, Technology, Business, Life.* Cambridge, MA: MIT Press, 2006.

Impact is heightened by what you leave out. Maeda is a former MIT professor who understands that good design is simple, and his advice is immediately applicable to communication and presentation.

Randy Olson, Dorie Barton, and Brian Palermo. *Connection: Hollywood Storytelling Meets Critical Thinking.* Los Angeles: Prairie Starfish Productions, 2013.

Olson is a PhD marine biologist who became a moviemaker. His book is intended for a broad audience, but his background as a scientist lets him tell the story of storytelling in ways that we as scientists understand. Much of storytelling is not equated with critical thinking—Olson tells us how to get a point across by engaging the hardwired aspects of human brains that absorb stories more readily than simple facts.

Kendall Haven. *Story Proof: The Science behind the Startling Power of Story.* Westport, CT: Libraries Unlimited, 2007.

Everyone tells us that stories are valuable ways to convey information, but Haven has investigated the effect in detail and gives us solid background for why this works and how we can engage it. He spends a lot of time with scientists, so his book is couched in terms that we understand.

Nancy Baron. *Escape from the Ivory Tower.* Washington: Island Press, 2010.

Baron brings the resources and experience of COMPASS, an organization dedicated to helping scientists talk to journalists and policymakers, to bear on helping you become a better speaker. With a focus on the press and journalism, she emphasizes making your message clear and understandable. Her "message box" approach is a great way to keep yourself on target when talking or being questioned.

Thomas S. Kuhn. *The Structure of Scientific Revolutions.* 50th anniversary ed. Chicago: University of Chicago Press, 2012.

Kuhn takes us on an extraordinarily deep dive into the science of science. Along the way, he exposes the fact that scientists are so deeply knowledgeable about their field that it changes the way they speak and, most importantly, how they think. If you want to have a much better understanding of how major changes occur in science, read this book, but give yourself time to digest it. It is not a book for the beach.

Marcel Mauss. *The Gift: The Form and Reason for Exchange in Archaic Societies.* New York: W. W. Norton, 1990.

Of all the books we read in preparing *Championing Science,* this was the most surprising. A deep dive into the structure and anthropology of societies before writing and commerce existed, it explains the basics of why the exchange of information is so valuable in working with decision makers and collaborators. The concept is more clearly exposed in other books, including Steven Covey's *The 7 Habits of Highly Effective People,* but the science of *why* it is important can be found here.

Robert Cialdini. *Influence: The Psychology of Persuasion.* Rev. ed. New York: Harper Business, 2006.

This is the classic popular description of the reasons that some people are persuasive and others fail. Although not aimed at scientific readers, it is very helpful to understand why many approaches that scientists regard as stupidly simplistic are in fact incredibly effective.

Steven Covey. *The 7 Habits of Highly Effective People.* 25th anniversary ed. New York: Simon and Schuster, 2013.

Begin with the end in mind. Covey's work is not aimed at presentations, but his advice on planning, relationships, and keeping your goals clearly in mind is relevant advice for the science champion.

Notes

1. BECOMING A CHAMPION

1. "The History of the Athanaeum," The Athenaeum, California Institute of Technology, accessed May 29, 2018, www.athenaeumcaltech.com/Default.aspx ?p=DynamicModule&pageid=342358&ssid=243574&vnf=1; Judith Goodstein, "History of Caltech," Nobelprize.org, June 29, 1998, www.nobelprize .org/nobel_prizes/themes/physics/goodstein; Judith Goodstein, *Millikan's School* (New York: W.W. Norton, 1996).

2. National Science Foundation, *Science, the Endless Frontier: A Report to the President by Vannevar Bush, Director of the Office of Scientific Research and Development* (Washington, DC: United States Government Printing Office, 1945; repr., New York: ACLS Humanities E-Book, 2008).

3. National Intelligence Council, *Global Trends 2030: Alternative Worlds,* (Washington, DC: National Intelligence Council, 2012), 3.

2. SELF-AWARE AND SELF-CORRECTING

1. Conversation with Roger Aines and Amy Aines at a meeting in Washington, DC, 2014. The staffer prefers to remain anonymous.

2. Quoted in Bill George, "Know Thyself: How to Develop Self-Awareness," *Psychology Today,* September 28, 2015, www.psychologytoday.com/blog /what-is-your-true-north/201509/know-thyself-how-develop-self-awareness.

3. Patrick Casement, *Learning from the Patient* (New York: Guilford Press, 1992).

3. EXTRACTING THE ESSENCE

1. Garr Reynolds, *Presentation Zen: Simple Ideas on Presentation Design and Delivery,* 2nd ed. (Berkeley, CA: New Riders Press, 2012), 56.

2. The Defense Threat Reduction Agency is the U.S. Department of Defense's official combat support agency for countering weapons of mass destruction. For more information, see the agency's website: www.dtra.mil/About.aspx.

3. Jay C. Davis and David A. Kay, "Iraq's Secret Nuclear Weapons Program," *Physics Today* 45, no. 7 (1992): 21–27, https://doi.org/10.1063/1.881312.

4. Center for Accelerator Mass Spectrometry, Lawrence Livermore National Laboratory (website), accessed June 3, 2018, https://cams.llnl.gov.

5. John Maeda, *The Laws of Simplicity: Design, Technology, Business, Life* (Cambridge, MA: MIT Press, 2006), chapter 10.

6. Roger Sessions, "How a 'Difficult' Composer Gets That Way," *New York Times*, January 8, 1950, www.nytimes.com/1950/01/08/archives/how-a-difficult-composer-gets-that-way-harpsichordist.html.

7. Alice Calaprice, ed., *The Ultimate Quotable Einstein* (Princeton: Princeton University Press, 2011), 384–385.

8. According to Wikiquotes, Einstein used this quote on two occasions: Einstein, "On the Method of Theoretical Physics," The Herbert Spencer Lecture (Oxford University, Oxford, June 10, 1933); and Einstein, "On the Method of Theoretical Physics," *Philosophy of Science* 1, no. 2 (April 1934): 165. See "Albert Einstein," Wikiquotes, last updated May 24, 2018, https://en.wikiquote.org/wiki/Albert_Einstein.

9. See the excellent tracing of the quote on the Quote Investigator website: Garson O'Toole, "Everything Should Be Made as Simple as Possible, but Not Simpler," Quote Investigator, May 13, 2011, www.quoteinvestigator.com/2011/05/13/einstein-simple.

10. Roger Aines and Amy Aines, "Influencing Decision Makers: From Julio Friedmann," Championing Science, March 4, 2014, www.championingscience.com/influencing-decision-makers.

11. This teacher's guide, for a game created for students to compare options, is available at the website of the Carbon Mitigation Initiative at Princeton University: Roberta Hotinski, "Stabilization Wedges: A Concept & Game," last updated March 16, 2015, http://cmi.princeton.edu/sites/default/files/wedges/pdfs/teachers_guide.pdf.

4. WHO'S LISTENING?

1. Suzanne Goldenberg, "Arizona Wildfire: Obama Lauds Elite Crew of 'Heroes' Who Died Tackling Blaze," *Guardian* (international edition), July 1, 2013, www.theguardian.com/world/2013/jul/01/arizona-wildfire-obama-heroes.

2. *Yarnell Hill Fire, June 30, 2013: Serious Accident Investigation Report,* Prescott Fire Department, September 23, 2013, available at www.hsdl.org/c/yarnell-hill-fire-serious-accident-investigation-report.

3. FirstNet-sponsored meeting in Denver, CO, May 21, 2013, transcript from Amy Aines's personal notes.

4. Ibid.

5. Steve Bohlen, conversation with Roger Aines and Amy Aines, Livermore, CA, September 8, 2013.

6. Susan Cain, *Quiet: The Power of Introverts in a World That Can't Stop Talking* (New York: Crown Publishing Group, 2012), Kindle.

7. "Quiet, Please: Unleashing 'The Power of Introverts,'" *All Things Considered*, NPR, January 30, 2012, www.npr.org/2012/01/30/145930229/quiet-please-unleashing-the-power-of-introverts.

8. Roger Aines and Amy Aines, "Influencing Decision Makers: From Julio Friedmann," Championing Science, March 4, 2014, www.championingscience.com/influencing-decision-makers.

9. Conversation with Roger Aines, 2013. The staffer prefers to remain anonymous.

10. Roger Aines and Amy Aines, "What Do Venture Capitalists Need to Hear from Scientists? A Conversation with Nancy Floyd, the Founder of Nth Power," Championing Science, June 24, 2017, www.championingscience.com/effective-communication-vcs-interview-nancy-floyd-nth-power.

5. WHY SCIENTISTS COMMUNICATE POORLY
OUTSIDE THEIR FIELD

Epigraph: Steven Pinker, *The Sense of Style: The Thinking Person's Guide to Writing in the 21st Century*, repr. (New York: Penguin, 2014) loc. 1047–1048 of 6647, Kindle.

1. For a short summary of Linnaeus's impact on scientific terminology, see www.ucmp.berkeley.edu/history/linnaeus.html.

2. Thomas S. Kuhn, *The Structure of Scientific Revolutions*, 50th anniversary ed. (Chicago: University of Chicago Press, 2012).

3. Ibid., 174.

4. Ian Hacking, introductory essay to Kuhn, *Structure of Scientific Revolutions*, xxiv.

5. Kuhn, *Structure of Scientific Revolutions*, 118.

6. Roger Aines and Amy Aines, "Getting to Yes in Biotech: Bill Young on Honing Your Message to Advance Science and Win Investor Support," Championing Science, April 12, 2016, www.championingscience.com/getting-yes-biotech-honing-message-advance-science-win-investor-support.

7. Dan M. Kahan, Ellen Peters, Maggie Wittlin, Paul Slovic, Lisa Larrimore Ouellette, Donald Braman, and Gregory Mandel, "The Polarizing Impact of Science Literacy and Numeracy on Perceived Climate Change Risks," *Nature Climate Change* 2 (2012): 732–735, doi:10.1038/nclimate1547.

8. Dan M. Kahan, "Making Climate-Science Communication Evidence-Based: All the Way Down," in *Culture, Politics and Climate Change: How Information Shapes our Common Future*, ed. Deserai A. Crow and Maxwell T. Boykoff (New York: Routledge, 2014), 203–220.

9. "If I Just Explain the Facts, They'll Get It, Right?" YouTube video, published by Global Weirding with Katharine Hayhoe on January 18, 2017, https://m.youtube.com/watch?v=nkMIjbDtdoo. See also Katharine Hayhoe, "When Facts Are Not Enough," *Science* 360, no. 6392 (2018): 943, doi:10.1126/science.aau2565.

10. Brendan Nyhan and Jason Reifler, "The Roles Of Information Deficits And Identity Threat In The Prevalence Of Misperceptions," February 24, 2017, www.dartmouth.edu/~nyhan/opening-political-mind.pdf.

11. The tenth version of this annual analysis can be found at "Levelized Cost of Energy Analysis 10.0," Lazard, December 15, 2016, www.lazard.com/perspective/levelized-cost-of-energy-analysis-100.

6. THE FIRST TWO MINUTES

1. Susan Weinschenk, *100 Things Every Presenter Needs to Know about People* (Berkeley, CA: New Riders, 2012), 171, Kindle.

2. Roger Aines and Amy Aines, "Steve Bohlen on Talking Science to Decision Makers: Explain Its Value to Society," Championing Science, January 21, 2014, www.championingscience.com/talking-science-decision-makers-explain-value-society.

3. Roger Aines and Amy Aines, "Influencing Decision Makers: From Julio Friedmann," Championing Science, March 4, 2014, www.championingscience.com/influencing-decision-makers.

4. "Scientists Detect Einstein's Gravitational Waves," YouTube video, published by AP Archive on November 17, 2016, www.youtube.com/watch?v=TMiVGeDi1yM.

5. Bruce Weber, "Swift and Slashing, Computer Topples Kasparov," *New York Times,* May 12, 1997, www.nytimes.com/1997/05/12/nyregion/swift-and-slashing-computer-topples-kasparov.html.

6. Ray Kurzweil, *The Age of Intelligent Machines* (Cambridge, MA: MIT Press, 1990), 133.

7. The University of Queensland maintains an excellent video repository of winning talks at www.threeminutethesis.uq.edu.au/watch-3mt.

8. "Three Minute Thesis (3MT) - Jenna Butler - 1st Place 2015," YouTube video, published by Western University on April 10, 2015, www.youtube.com/watch?v=GydZ1cRKWLg.

7. CRAFTING KEY MESSAGES AND NARRATIVES

1. Roger Aines and Amy Aines, "Getting to Yes in Biotech: Bill Young on Honing Your Message to Advance Science and Win Investor Support," Championing Science, April 12, 2016, www.championingscience.com/getting-yes-biotech-honing-message-advance-science-win-investor-support.

2. Roger Aines and Amy Aines, "What Do Venture Capitalists Need to Hear from Scientists? A Conversation with Nancy Floyd, the Founder of Nth Power," Championing Science, June 24, 2017, www.championingscience.com/effective-communication-vcs-interview-nancy-floyd-nth-power.

3. Roger Aines and Amy Aines, "Influencing Decision Makers: From Julio Friedmann," Championing Science, March 4, 2014, www.championingscience.com/influencing-decision-makers.

4. "UMN 3MT Dustin Chernick," YouTube video, published by UMN Health Science on June 2, 2016, www.youtube.com/watch?reload=9&v=hN7UtB-P004.

5. "ANU 3MT 2015: Suryashree Aniyan," YouTube video, published by ANU TV on October 8, 2015, www.youtube.com/watch?v=KsaLz71A_FA 9/16/2015.

6. Steven Covey, *The 7 Habits of Highly Effective People,* 25th anniversary ed. (New York: Simon and Schuster, 2013), 95.

7. Amy Aines and Roger Aines, "Jay Davis on Accelerator Mass Spectrometry's Impact on Medicine and the Importance of a Single Page of Text," Championing Science, February 4, 2014, www.championingscience.com/accelerator-mass-spectrometrys-impact-medicine-importance-good-pitch.

8. FirstNet-sponsored meeting in Denver, CO, May 21, 2013, transcript from Amy Aines's personal notes.

9. Eric Blackman, "Closed TBI and Its Protection: A Physics Perspective," accessed July 8, 2018, www.pas.rochester.edu/~blackman/helmetsTBI2014 rochester.pdf.

10. Kendall Haven, *Story Proof: The Science behind the Startling Power of Story* (Westport, CT: Libraries Unlimited, 2007).

11. Roger Aines and Amy Aines, "Dr. Richard Friedman: Champion of Heart Health," Championing Science, February 27, 2014, www.championing science.com/dr-richard-friedman-champion-heart-health.

12. Lisa Cron, *Wired for Story: The Writer's Guide to Using Brain Science to Hook Readers from the Very First Sentence* (New York: Ten Speed Press, 2012).

13. Roger Aines and Amy Aines, "Getting to Yes in Biotech," www .championingscience.com/getting-yes-biotech-honing-message-advance-science-win-investor-support.

14. Robert Bazell, *Her-2: The Making of Herceptin, a Revolutionary Treatment for Breast Cancer* (New York: Random House, 1998).

15. Randy Olson, Dorie Barton, and Brian Palermo, *Connection: Hollywood Storytelling Meets Critical Thinking* (Los Angeles: Prairie Starfish Productions, 2013).

8. THE POWER OF LANGUAGE

1. ORIC Pharmaceuticals homepage, accessed May 25, 2018, www .oricpharma.com.

2. "Cancer: The Emperor of All Maladies," PBS, accessed May 25, 2018, www.pbs.org/show/story-cancer-emperor-all-maladies; emphasis added.

3. "Climate Change," FrameWorks Institute, March 21, 2014, www .frameworksinstitute.org/climate-change-and-the-ocean.html.

4. William Strunk and E. B. White, *The Elements of Style,* 4th ed. (Needham Heights, MA: Allyn and Bacon, 2000).

5. Roger Aines and Amy Aines, "What Do Venture Capitalists Need to Hear from Scientists? A Conversation with Nancy Floyd, the Founder of Nth Power," Championing Science, June 24, 2017, www.championingscience.com/effective-communication-vcs-interview-nancy-floyd-nth-power.

6. Rob Socolow, conversation with Amy and Roger Aines, 2014.

7. Arnold Samuelson, *With Hemingway: A Year in Key West and Cuba* (New York: Random House, 1984). This quote is cited and described on the

Quote Investigator website: Garson O'Toole, "The First Draft of Anything Is Shit," Quote Investigator, September 20, 2015, www.quoteinvestigator.com /2015/09/20/draft.

8. Steven Pinker, *The Sense of Style: The Thinking Person's Guide to Writing in the 21st Century,* repr. (New York: Penguin, 2014).

9. Francis-Noël Thomas and Mark Turner, cited in Pinker, *Sense of Style,* 28.

10. Francis-Noël Thomas and Mark Turner, *Clear and Simple as the Truth: Writing Classic Prose,* 2nd ed. (Princeton, NJ: Princeton University Press, 2011), 4.

11. Pinker, *Sense of Style,* 56.

12. Ibid., 55.

13. Ibid., 55–56.

14. Ibid., 56.

9. DESIGNING EFFECTIVE VISUALS

Epigraph: Amy Aines and Roger Aines, "The Art of Making Your Audience Comfortable: A Conversation about Structure, Style and Strategy with Joshua White," Championing Science, February 13, 2014, www.championingscience .com/art-making-audience-comfortable-structure-style-strategy-2.

1. A high-quality video of this can be found at www.youtube.com /watch?v=ZOzoLdfWyKw. The O-ring demonstration begins at 2:50.

2. William L. Ellsworth, "Injection-Induced Earthquakes" *Science* 341, 1225942 (2013). DOI: 10.1126/science.1225942

3. Edward R. Tufte, *The Visual Display of Quantitative Information,* 2nd ed. (Cheshire, CT: Graphics Press, 2001); Tufte, *Envisioning Information* (Cheshire, CT: Graphics Press, 1990); Tufte, *Visual Explanations: Images and Quantities, Evidence and Narrative* (Cheshire, CT: Graphics Press, 1997); Tufte, *Beautiful Evidence* (Cheshire, CT: Graphics Press, 1997).

4. Felice C. Frankel, *Envisioning Science: The Design and Craft of the Science Image* (Cambridge, MA: MIT Press, 2002); Felice C. Frankel and Angela H. DePace, *Visual Strategies* (New Haven, CT: Yale University Press, 2012).

5. Steve Calechman, "Making More than Pretty Pictures," MIT News, January 29, 2015, http://news.mit.edu/2015/felice-frankel-making-more-than-pretty-pictures-0129.

6. Garr Reynolds, *Presentation Zen: Simple Ideas on Presentation Design and Delivery,* 2nd ed. (Berkeley, CA: New Riders Press, 2012).

7. Ibid., 120–121.

8. Garr Reynolds, quoted in Nancy Duarte, *Slide:ology The Art and Science of Creating Great Presentations* (Sebastopol, CA: O'Reilly Media, 2008), 109.

9. "Al Gore: An Inconvenient Truth, the Climate Change Presentation," Duarte, accessed July 8, 2018, www.duarte.com/portfolio/al-gore.

10. Susan Weinschenk, *100 Things Every Presenter Needs to Know about People* (Berkeley, CA: New Riders, 2012).

11. "Here's The RIGHT Way to Show Your Company Logo on Your Slides—Be Distinctive (Not Dismissive)," Remote Possibilities (blog), February

12, 2013, www.remotepossibilities.wordpress.com/2013/02/12/the-right-way-to-show-your-company-logo-on-your-slides-be-distinctive.

10. IMPROVING YOUR SPEAKING SKILLS

1. "UMN 3MT Dustin Chernick," YouTube video, published by UMN Health Science on June 2, 2016, www.youtube.com/watch?v=hN7UtB-P004.

11. INFLUENCE AND PERSUASION

Epigraph: Lesley Brown, ed., *New Shorter Oxford English Dictionary,* Thumb Index ed. (New York: Oxford University Press, 1993), s.vv. "influence," "persuasion."

1. Randy Olson, Dorie Barton, and Brian Palermo, *Connection: Hollywood Storytelling Meets Critical Thinking* (Los Angeles: Prairie Starfish Productions, 2013).

2. Daniel Kahneman, *Thinking, Fast and Slow* (New York: Farrar, Straus and Giroux, 2011).

3. Ibid., 20.

4. Daniel Kahneman, "Thinking We Know," (lecture presented at The Science of Science Communication, Arthur M. Sackler Colloquia of the National Academy of Sciences, Washington, DC, May 21–22, 2012, National Academy of Sciences), http://events.tvworldwide.com/Events/NAS120521.

5. Olson gives an excellent summary of Kahneman's lecture; see Olson, Barton, and Palermo, *Connection,* 50. See also Bud Ward, "Revisiting National Academy's 'Sackler Colloquium,'" pt. 1, *Yale Climate Connections,* June 6, 2012, www.yaleclimateconnections.org/2012/06/revisiting-national-academys-sackler-colloquium-part-i.

6. Ibid.

7. "How to Build a Brain," YouTube video, published by ANU TV on October 20, 2014, www.youtube.com/watch?v=yTkSAceGenw&list=PLdnogiYPTOk3y2B2KL2jtS6mHvMPFdj7u.

8. Roger Aines and Amy Aines, "What Do Venture Capitalists Need to Hear from Scientists? A Conversation with Nancy Floyd, the Founder of Nth Power," Championing Science, June 24, 2017, www.championingscience.com/effective-communication-vcs-interview-nancy-floyd-nth-power.

9. Marcel Mauss, *The Gift: The Form and Reason for Exchange in Archaic Societies,* translated by W.D. Halls, foreword by Mary Douglas (New York: W.W. Norton, 1990).

10. "2016 CEHD Three Minute Thesis Competition—First Prize Winner Michelle Brown," YouTube video, posted by UMCEHD on April 1, 2016, www.youtube.com/watch?v=Elt_oP8_iZE.

11. Amy Cuddy, *Presence: Bringing Your Boldest Self to Your Biggest Challenges* (New York: Little, Brown, 2015), 25.

12. Roger Aines and Amy Aines, "Getting to Yes in Biotech: Bill Young on Honing Your Message to Advance Science and Win Investor Support," Championing Science, April 12, 2016, www.championingscience.com/getting-yes-biotech-honing-message-advance-science-win-investor-support.

13. Roger Aines and Amy Aines, "Steve Bohlen on Talking Science to Decision Makers: Explain Its Value to Society," Championing Science, January 21, 2014, www.championingscience.com/talking-science-decision-makers-explain-value-society.

14. Jay A. Conger, "The Necessary Art of Persuasion," *Harvard Business Review* 76, no. 3 (May-June 1998): 84–95, www.hbr.org/1998/05/the-necessary-art-of-persuasion.

15. "Examples of Ethos, Logos, and Pathos," YourDictionary, accessed June 8, 2018, http://examples.yourdictionary.com/examples-of-ethos-logos-and-pathos.html#DZ5OvlXhoJKWxpyL.99.

16. Amy Aines and Roger Aines, "The Art of Making Your Audience Comfortable: A Conversation about Structure, Style and Strategy with Joshua White," Championing Science, February 13, 2014, www.championingscience.com/art-making-audience-comfortable-structure-style-strategy-2.

12. MANAGING YOUR EMOTIONS

1. Nick Morgan, "Why We Fear Public Speaking and How to Overcome It," *Forbes,* March 30, 2011, www.forbes.com/sites/nickmorgan/2011/03/30/why-we-fear-public-speaking-and-how-to-overcome-it.

2. Amy Cuddy, *Presence: Bringing Your Boldest Self to Your Biggest Challenges* (New York: Little, Brown, 2015), 20–21, Kindle.

3. Susan Cain, *Quiet: The Power of Introverts in a World That Can't Stop Talking* (New York: Crown Publishing Group, 2012), 129, Kindle.

4. Susan Cain, "The Power of Introverts," TED, video, February 2012, www.ted.com/talks/susan_cain_the_power_of_introverts.

5. Adam Hajo and Adam D. Galinsky, "Enclothed Cognition," *Journal of Experimental Social Psychology* 48 (2012): 918–925, http://dx.doi.org/10.1016/j.jesp.2012.02.008.

6. Cuddy, *Presence,* 18, Kindle.

7. Neck, C.P., Nouri, H., Godwin, J.L. (2003). How self-leadership affects the goal-setting process. Human Resource Management Review, 13(4): 691–707.

8. Scott Williams, "Head Games: The Use of Mental Rehearsal to Improve Performance," newsletter, accessed June 14, 2018, www.wright.edu/~scott.williams/LeaderLetter/rehearsal.htm. Procedure based on C.C. Manz and C.P. Neck, *Mastering Self-Leadership: Empowering Yourself for Personal Excellence,* 2nd ed. (Upper Saddle River, NJ: Prentice Hall, 1999), 70–71.

9. Cuddy, *Presence,* 88–89, Kindle.

10. Ibid., 103, Kindle.

11. Neil Gaiman's Tumblr blog, post dated May 12, 2017, www.neil-gaiman.tumblr.com/post/160603396711/hi-i-read-that-youve-dealt-with-with-impostor.

13. TRANSLATIONS, TEMPLATES, AND WHITE PAPERS

Epigraph: Roger Aines and Amy Aines, "Jay Davis on Accelerator Mass Spectrometry's Impact on Medicine and the Importance of a Single Page of Text,"

Championing Science, February 4, 2014, www.championingscience.com
/accelerator-mass-spectrometrys-impact-medicine-importance-good-pitch.

1. Garr Reynolds, *Presentation Zen: Simple Ideas on Presentation Design and Delivery* (Berkeley, CA: Pearson Education, 2011), 158–159.

14. STRATEGIES FOR CREATING SUCCESSFUL RELATIONSHIPS WITH SPONSORS

1. Roger Aines and Amy Aines, "Influencing Decision Makers: From Julio Friedmann," Championing Science, March 4, 2014, www.championingscience .com/influencing-decision-makers.

16. HIGH-IMPACT EXAMPLES OF CHAMPIONING FOR A CAUSE

1. The famous efforts to convince Roosevelt to investigate nuclear weapons were conducted later under the successors to the National Defense Research Committee.

2. G. Pascal Zachary, *The Endless Frontier: Vannevar Bush, the Engineer of the American Century* (Cambridge, MA: MIT Press, 1999).

3. National Science Foundation, *Science, the Endless Frontier: A Report to the President by Vannevar Bush, Director of the Office of Scientific Research and Development* (Washington, DC: United States Government Printing Office, 1945; repr., New York: ACLS Humanities E-Book, 2008).

4. Zachary, *Endless Frontier*, 112.

5. Ibid., 102.

6. Ibid., 38.

7. Ibid., 105.

8. Ibid., 36.

9. Ibid., 108.

10. "Scientists Detect Einstein's Gravitational Waves," YouTube video, published by AP Archive on November 17, 2016, www.youtube.com/watch?v= TMiVGeDi1yM.

11. Davide Castelvecchi, "Hunt for Cosmic Waves to Resume: Upgraded LIGO Detectors Will Improve Chances of Finding Ripples in Space-Time," *Nature* 525, no. 7569 (2015): 301.

12. Rainer Weiss, interviewed by Shirley K. Cohen, Pasadena, CA, May 10, 2000, Oral History Project, California Institute of Technology Archives, http:// oralhistories.library.caltech.edu/183/1/Weiss_OHO.pdf.

13. Caltech Oral History Project, http://oralhistories.library.caltech.edu; Janna Levin, *Black Hole Blues and Other Songs from Outer Space* (New York: Knopf, 2016).

14. Rainer Weiss interview, Caltech Oral History Project, 17.

15. Levine, *Black Hole Blues*, 17.

16. That student was Bill Press, who went on to become the youngest tenured faculty member at Harvard, the deputy director of the Los Alamos National Laboratory, and the president of the American Association for the Advancement of Science.

17. Rainer Weiss, interview, Caltech Oral History Project, 24.

18. Ibid., 25.

19. Quoted in Robert Buderi, "Going after Gravity: How a High-Risk Project Got Funded," *Scientist Magazine,* September 19, 1988, www.the-scientist.com /news/going-after-gravity-how-a-high-risk-project-got-funded-62645.

20. Ibid.

21. Rainer Weiss, interview, Caltech Oral History Project, 27.

22. Ibid.

23. Ibid., 30.

24. National Research Council, "Appendix C: Histories of Projects Funded by NSF," in *Setting Priorities for Large Research Facility Projects Supported by the National Science Foundation* (Washington, DC: National Academies Press, 2004), doi: 10.17226/10895.

25. Buderi, "Going after Gravity."

26. Ranier Weiss, interview, Caltech Oral History Project, 28.

27. Charles W. Peck, interviewed by Shirley K. Cohen, Pasadena, CA, October 1, 8, 15, and 30 and November 12, 2003, Oral History Project, California Institute of Technology Archives, http://oralhistories.library.caltech.edu/237/1 /Peck%2C%20Charles_OHO.pdf, 71.

28. National Research Council, "Appendix C: Histories of Projects Funded by NSF."

29. Charles W. Peck, interview, Oral History Project, 72.

30. Leon T. Silver, interviewed by Shirley K. Cohen, Pasadena, CA, December 12 and 20, 1994, January 9, 16, and 25, 1995, and February 23, 2000, Oral History Project, California Institute of Technology Archives, http://oralhistories .library.caltech.edu/137/1/Silver_OHO.pdf, 78.

31. Steve Bohlen, conversation with Amy Aines and Roger Aines, 2015.

32. Roger Aines and Amy Aines, "Jay Davis on Accelerator Mass Spectrometry's Impact on Medicine and the Importance of a Single Page of Text," Championing Science, February 4, 2014, www.championingscience.com/accelerator-mass-spectrometrys-impact-medicine-importance-good-pitch.

33. Graham Moore, *The Last Days of Night: A Novel* (New York: Random House), 31, Kindle.

Index

acknowledgements, 33–35, 105, 121, 122

acronyms, 9, 37, 43, 44–45, 90, 201, 213

actions: benefits of taking, 79; calls to, 20, 143, 167, 206; championing science, 8–9; congressional, 38; inhibitions to, 134, 149; Roosevelt's, 208; self-awareness and, 11. *See also* enabling audience to act; inspiring audience to act

active voice, 93–94

adrenaline, 160, 163, 164, 167

Age of Intelligent Machines, The, 61

AirTouch Communications, 71, 151

Althouse, Bill, 218

Alzheimer's, 69, 75, 129, 150

ammonia, 195–97

analogies, 69, 77, 85–87, 152. *See also* iconic analogies

analysis, 15, 21, 27, 100, 165

And-But-Therefore model, 83, 100, 102, 104, 199

applicability, 178–79

Aristotle, 146, 150

arriving early, 30–31

Art of Rhetoric, The, 146

ask (the): audience's constraints, 143; closing, 168; compelling audience to agree to, 72, 90; defining, 195; definitive, 135; deliverables, 141, 143; enabling audience to act, 9; five slide approach, 17–18, 20–22, 200, 206; key messages, 70, 168; preliminary

proposal, 153; time management, 186–88; white papers, 182–85

attire, 63–64, 162–63, 167, 206

attribution, 122, 182

audiences: analogies for, 77; and the ask, 20; attention span, 15; attention to attire, 64; and biographies, 204; boring, 21, 100, 178; captivating, 92–93; challenging, 19, 54; connecting with, 20, 30–31, 34, 35, 65; crediting collaborators, 29, 34, 146; directing discussion, 17; earning trust, 137, 138–41, 150, 181; engaging, 35, 52, 54; exciting, 57, 61, 137–38, 178, 181, 205; expectations, 52; failure with, 13, 16, 22; gestures and, 129–30, 147; impacting, 22, 45; impromptu, 40–41; influencing, 9; insulting, 104, 181; and key messages, 16, 72–74, 77; learning styles, 36–37; listening to, 12, 13; mixed, 121, 122*fig*; moving in front of, 63, 129–30, 147, 160, 162; multilingual, 175–76; multiple, 72–74; nontechnical, 7, 62; opening lines for, 58–59, 61, 62; paradigms, 50; phenotypes, 29–30; practice, 169, 206; preparatory research, 9, 27–28, 29–30, 34, 40, 49, 77, 193–94, 203, 206; Q&A sessions with, 131–35; simplifying visuals for, 26*fig*; stage fright, 160–65; stories and, 80–81, 83, 104; talking to,